P9-AQE-674

PROTECTING ENGINEERING IDEAS
& INVENTIONS

Ramon D. Foltz
Intellectual Property Lawyer

Thomas A. Penn
Engineering & Management Consultant

PENN INSTITUTE
P.O. BOX 41016 • CLEVELAND, OHIO 44141

216-237-2345
800-426-7495

ISBN 0-944606-01-6 (Paper Back Edition)
ISBN 0-944606-02-4 (Hard Back Edition)

Copyright 1987, 1988 Penn Institute, Inc.
All Rights Reserved
Appendices A and D are derived from U.S. Government publications.

KF
3114.8
.E54
F66
1988

DISCLAIMER

While the information in this book, *Protecting Engineering Ideas & Inventions*, is believed to be current and accurate, and is intended to help an engineering department do their part in protecting the company's investment in engineering, no writing can replace sound judgement or make decisions. Penn Institute, Inc. cannot assume any responsibility for any decisions made or actions taken based on the information in this book, *Protecting Engineering Ideas & Inventions*, or any other such materials provided by Penn Institute, Inc.

Further, *Protecting Engineering Ideas & Inventions*, is sold with the understanding that Penn Institute, Inc. is not engaged in rendering legal or other professional services or advice. If legal advice or other expert assistance is required, be sure to obtain the personal services of a competent professional.

ABOUT THE AUTHORS

This book could not have been written by someone whose only expertise is the legalities of protecting engineering ideas and inventions. Rather, the writing of this book also required knowing exactly what an engineering department needs to know about such protection, and how the information should be explained and organized so that the department personnel find it useful and convenient to use. Thus, this book was written jointly by R.D. Foltz, a corporate intellectual property lawyer, and T.A. Penn, a seasoned engineering manager.

Since 1965, R.D. Foltz has been an intellectual property lawyer for TRW Inc. and Eaton Corporation. Every day, Mr. Foltz works closely with engineering departments. Mr. Foltz knows how to protect intellectual property and what role **engineering personnel** need to play in such protection.

On the other hand, since 1960 T.A. Penn held positions in engineering and engineering management in major firms such as RCA, General Electric, and Gould, Inc. In 1979 Mr. Penn founded Penn Institute, Inc., an engineering consulting company. Mr. Penn has lived the engineer's life -- he knows how legal information must be presented so that it's easy to read and dovetails with the engineer's day-to-day activities.

Mr. Foltz and Mr. Penn combined their experience and knowledge to put between two covers what engineering personnel should know about protecting their company's engineering investment via patents, copyrights, trademarks, trade secrets, and secrecy agreements; the best way to work with intellectual property lawyers and outside consultants; and how to handle outsiders ideas.

TABLE OF CONTENTS

FIGURES

PART I. INTRODUCTION

One responsibility an engineering department has is to provide their company with competitively superior products. Simply stated, an engineering department accomplishes this by generating ideas, improving on those ideas until they become inventions, and ultimately developing these inventions into profitable products. A company has nothing to gain -- but has a lot to lose -- if competitors are free to copy these engineering concepts before the company has a chance to profit from them. When a company says "We had the idea but the XYZ company stole it and came up with the product!" it probably means the idea-generating company will end up with only a "me too" product instead of an exclusive one -- and, in effect, the return on the company's engineering investment will be reduced, if not lost completely.

This book was written specifically to help engineering departments do their part in using the intellectual property laws to protect a company's engineering work from being legally copied. To this end, this book will allow engineering personnel to better understand:

(1)   Intellectual property concepts, and

(2)   The forms of legal protection important to engineering work (namely, patents, copyrights, trademarks, and trade secrets).

The book also provides guidance on:

(3)   Secrecy agreements with outsiders,

(4)   The best way to work with outsiders' ideas and outside consultants, and

(5)   If the company doesn't have legal help, when to involve or hire lawyers, what kind of lawyer is needed and when, and how to work with lawyers and what to expect from them.

It's important to realize that even if an engineering department has access to lawyers, the engineering personnel must do their part in protecting engineering work or the company may not even have the opportunity to get a patent or other form of protection.

1

Since your primary expertise is engineering, you probably learned what you know about protecting engineering information from conversations with lawyers, or perhaps you have read a law book or two. Most likely, the books you read were either written for lawyers or not presented in the fashion in which engineering personnel need the information. You will find this book to be completely different in this regard -- its information dovetails with day-to-day engineering activities and one <u>does</u> <u>not</u> have to be a lawyer to understand it.

I.1 The Concept Of Intellectual Property

Engineering ideas and inventions are considered to be forms of "intellectual property." Loosely defined, "intellectual property" is "products of the mind." Further:

*   Patents, copyrights, trademarks, and trade secrets are methods of <u>protecting</u> intellectual property.

*   Laws which pertain to patents, copyrights, trademarks, trade secrets, and secrecy agreements are called "intellectual property laws."

*   Lawyers who assist in the protection of ideas, inventions, and products are called "intellectual property lawyers," "patent lawyers," "patent attorneys," or for trademark work, "trademark lawyers."

I.2 Concluding Points

(1)   There is no need to read this book from cover to cover at one sitting. It's uniquely organized so that you need only read the portion which is of interest in a given situation.

(2)   Most often, legal language sounds confusing because our lawmakers must account for as many situations as they can conceive. In this book we translated and tailored this legal language so that engineering people would find this book easy to read.

(3)   You will find that the legal subject matter and procedures covered here are more straightforward than you once might have thought.

(4)   You will not be able to start your own law practice just because you read this book but you will have the information you need to effectively carry out your intellectual property protection responsibilities.

(5)   United States' intellectual property laws are changing constantly as technology advances and new cases are brought before the courts. To the best of our knowledge, the information in this book is accurate as of May 1, 1988 but with each passing day some small portion of it may become obsolete.

In short, this book was written especially to help engineering personnel handle the engineering-related patent activities and to deal with copyrights, trademarks, trade secrets, and secrecy agreements. It also advises engineering personnel on the best way to work with the proper lawyers and outsiders, including outside consultants. And -- of equal importance to the busy engineer -- this book is easy to read.

PART II. ABOUT LAWYERS

This book was written to help you, an engineering person, do your part in protecting your company's engineering ideas and inventions from being copied and used by your competitors. However, you will still need assistance from qualified legal professionals. If your company already has lawyers then the job of finding help is obviously made easier. If your company doesn't have that luxury, engineering personnel must know what kind of help to seek. To this end, this section provides some general information about the legal profession as far as the protection of intellectual property is concerned.

II.1 Lawyers And Intellectual Property Lawyers

Lawyers are trained to read, understand, interpret, and apply the law; and are licensed by a state or the District of Columbia to represent others before various courts and administrative agencies. Lawyers often specialize their practice, e.g. real estate, banking, and divorce. Lawyers who specialize in patents, copyrights, trademarks, trade secrets, and related areas are generally known as "intellectual property lawyers;" and while there are approximately 600,000 lawyers in the U.S., only a relatively few fall into the intellectual property lawyer category. Further, when it comes to U.S. patent work, only about 12,000 persons are even <u>allowed</u> to provide some patent services. (This restriction will be explained in section II.2 below.)

It's vitally important to realize that the <u>quality</u> of the intellectual property protection your company will obtain is highly dependent on the <u>skill</u> of the intellectual property lawyer. Therefore, if your company doesn't already have the legal help you need, be sure to thoroughly check out an outsider before you contract for help. Interview him or her, ask for references, and speak to those references.

II.2 Patent Legal Experts

There are two types of intellectual property professionals that specialize in patents -- *patent lawyers* and *patent agents*.

First, *patent lawyers* are lawyers because they:

5

(a)   have a law degree, and

(b)   have passed at least one state bar examination.

Secondly, patent lawyers are generally qualified to work with patents because they:

(c)   have an engineering degree or have otherwise demonstrated competence in a technical field,

(d)   have passed a special examination given by the U.S. Patent and Trademark Office, and therefore,

(e)   are "registered" by the U.S. Patent and Trademark Office, and are allowed to practice before the U.S. Patent and Trademark Office.

In addition to providing patent services, patent lawyers (who oftentimes are referred to in the broad term "intellectual property lawyers") usually have the experience and education required to give advice and carry out legal activities for copyrights, trademarks, trade secrets, and related legal matters.

Patent agents are usually engineers or otherwise technically trained persons, have passed the U.S. Patent and Trademark Office examination, and therefore are registered to practice before the U.S. Patent and Trademark Office. However, patent agents are not lawyers; but like patent lawyers, patent agents usually have the experience necessary to handle copyrights and trademarks.

Despite the distinction of having or not having a law degree, there should be no difference between a skilled patent lawyer and a skilled patent agent when it comes to preparing and prosecuting a patent application in the U.S. Patent and Trademark Office.

> **Note:** In this book, we generally use the term "patent lawyers" to refer to both patent lawyers and patent agents.

Patent law permits an inventor to "act for himself" in obtaining a patent. This means the inventor can do everything himself if he so chooses, or anyone can help the inventor <u>write</u>

6

the patent application. However, if the inventor wants someone to represent him in corresponding with the Patent and Trademark Office, that representative <u>must</u> <u>be</u> either a <u>registered</u> patent lawyer or a <u>registered</u> patent agent.

The Patent and Trademark Office will not recommend a specific patent lawyer but does publish a directory, arranged by state and city, of the more than 12,000 registered patent lawyers and agents. This directory is entitled *Attorneys and Agents Registered to Practice Before the U.S. Patent and Trademark Office* and is available for $17.00 from:

> Superintendent of Documents
> U.S. Government Printing Office
> Washington, DC 20402

In addition to knowing the technical subject matter involved, applying for a patent also requires knowing the patent laws and procedures -- and it's very easy to get lost in either of these legal arenas. Therefore, while there is much engineering personnel must do in the process of getting a patent, we <u>strongly</u> recommend that the department also enlists the services of a registered patent lawyer. By not using a patent lawyer, you may be putting your company at a great disadvantage. For example, if you do not apply for broad enough protection the patent may not adequately or fully protect the invention. Conversely, if you apply for protection which is too broad your patent may be invalid. Hence, while patent lawyers may appear to be expensive, obtaining patent protection for a commercially valuable invention more than justifies the cost of working with an experienced and competent patent lawyer. If the company does not have such a person, someone needs to identify, interview, and select a good patent lawyer from the more than 12,000 patent lawyers available.

II.3 Copyright Legal Experts

Copyright law is a subspecialty within intellectual property law, but lawyers that practice copyright law do not have to be patent lawyers nor do they have to be registered with any government agency. Most patent lawyers have experience in copyright law and are able to provide related legal services. If a copyright situation is complex it's best to work with a lawyer (patent lawyer or not) who <u>specializes</u> in copyright matters.

II.4 Trademark Legal Experts

Trademark law is also a subspecialty within intellectual property law. Usually the legal experts here are patent and general lawyers who are skilled in the commercial aspects of business such as advertising, franchising, marketing, and sales.

II.5 Trade Secret And Secrecy Agreement Legal Experts

Any lawyer experienced in drafting agreements should be able to provide assistance with trade secrets and secrecy agreements. However, since such agreements often require knowledge of engineering and technology, we believe that intellectual property lawyers should be consulted in connection with trade secrets and engineering-related secrecy agreements.

II.6 Summary

Engineering personnel must handle some of the aspects of patents, copyrights, trade secrets, trademarks, and secrecy agreements on their own, but often will need the advice and help of qualified legal experts. However, be careful if you have to go outside of your company for help.

If your problem deals with copyrights, trademarks, trade secrets, or secrecy agreements -- and even though any lawyer can legally work in these areas -- your department should strive to find a lawyer, patent lawyer, or patent agent who is an expert in the specific subject.

If your problem is patent-related, make sure the person is not only skilled in patents but is registered with the Patent and Trademark Office. Only registered patent lawyers or agents can represent an inventor before the Patent and Trademark Office. Further, we strongly recommend that engineering personal always use the services of a skilled and registered patent lawyer or agent even though patent law allows an inventor to do his own patent work.

PART III. OVERVIEW OF UNITED STATES PATENTS

Of all of the intellectual property protection methods included in this book, patents are of the greatest importance to an engineer.

III.1 Why Do Patents Exist?

A patent is a government-granted and secured legal right to prevent others from practicing (i.e. making, using, or selling) an invention; and is also a form of personal property which can be licensed, sold, mortgaged, willed, assigned, or inherited. U.S. patents are granted by the U.S. Patent and Trademark Office. However, before a patent is granted, the inventor must file a patent application with the Patent and Trademark Office and the invention must meet the requirements established by law. Federal patent laws have pre-empted the states from having patent laws, which means the only patent laws are federal laws and there are no such things as state patent laws or state patents.

What's an "invention?"

To help in understanding patents and patent law it's best to view an "invention" in terms of one of its definitions, i.e. "an act of mental creation or organization." This means that a concept is an "invention" even if the concept is not useful, new, or novel. However -- as discussed throughout this book -- strict requirements must be met before an "invention" becomes a "patentable invention." (See Figure 4 in Part V, Chapter 3.)

Starting from the date it's issued, a patent is in effect for a term of 14 or 17 years, depending on the type of patent; and under normal circumstances a patent can not be renewed. During the term of a patent, the patent is presumed to be valid and enforceable.

U.S. patent laws are designed to "promote the progress of the useful arts." The theory is that, by encouraging inventors to reveal the details of their inventions, the patent laws will increase the technical knowledge which is publicly available in the United States. It works this way: The patent application reveals the details of the invention. The Patent and

Trademark Office then examines the application and, if the invention is patentable, they issue and publish a patent containing those details and provide the patent owner with patent protection. This protection gives the patent owner a legally enforceable right to **PREVENT** others from making, using, or selling the technical information, product, process, or technology which is claimed in the patent, but the patent does <u>not</u> give the patent owner any right to make, use, or sell his own invention. (See Section III.6, "Patent Rights" in this part and Part V, Chapter 7, "Avoiding Patent Infringement.")

III.2 <u>What Can Be Patented?</u>

The law defines what can be patented and the types of patents as follows:

(1)   A "utility" patent may be obtained on any new, useful, and unobvious process (primarily an industrial or technical process); machine; article of manufacture; or composition of matter (generally chemical compounds, formulas, and like); including any new, useful, and unobvious modification or improvement of prior technology. By far, a utility patent is the most popular type of patent.

(2)   A "design" patent may be obtained on any new, original, and unobvious ornamental design for an article of manufacture.

        Design patents, which can be obtained on designs such as a new auto body, may not turn out to be valuable because a commercially similar design can often be made without infringing the patent.

(3)   A "plant" patent may be obtained on any distinct and new variety of plant (except plants which are tuber propagated or plants found in an uncultivated state) which has been asexually produced by the inventor.

        Since engineering work seldom involves plants -- and there are only a very few plant patents each year -- let's wrap up plant patents by saying that you can't patent a potato but you stand a chance with a new variety of fruit tree.

In general, you can patent (and therefore protect) any original or improved: product or ornamental design; manufacturing process; mechanical, electric and electronic device;

chemical and metallurgical composition; or asexually produced life form. Now, let's look at what is <u>not</u> patentable:

(1)   You can't patent anything which is not "useful."

(2)   You can't patent anything which is not new. That is, it must be different from (but not necessarily better than) the "prior art." "Prior art" is the publicly known cumulative technical experience and knowledge of every person who has ever lived (see Section III.5, "Prior Art: A Major Factor Of Patentability" in this part.)

(3)   Although it may be new and useful, you can't patent anything which is an obvious modification of the "prior art" or is merely the result of applying ordinary mechanical skill such as a craftsman would possess (e.g. using aluminum window frames instead of wood frames). See Part V, Chapter 3, "How To Evaluate Obviousness: A Major Factor Of Patentability."

(4)   An "idea" cannot be patented; rather, the idea must first be "reduced to practice." (See Part V, Chapter 1, "How And Why To Record The Important Dates.") Then, in the eyes of the law, the "idea" becomes an "invention" and the <u>invention</u> may be eligible for patent protection. The law allows for an idea to be reduced to practice in either one of two ways:

(a)   having a working model, or

(b)   filing a patent application which describes the invention in complete enough terms so that a person of ordinary skill in the art can make and use the invention.

In other words, only "inventions" are patentable and ideas become inventions when it's demonstrated that the idea results in a useful process, machine, article of manufacture, or composition of matter.

(5)   An inoperable device cannot be patented. To put it another way, if it's built and it doesn't work, it's still not an "invention" but only an "idea."

(6)   You can't patent a method of doing business, such as developing a new parts numbering system; however, any new equipment that is used in generating that numbering system might be patentable.

(7)   Printed matter cannot be patented (this is covered by copyright law -- See Part VI, "Copyrights").

III.3 To Whom (And When) A Patent May Be Granted

Patents are not just granted to anyone who says they have an invention. Patents are only granted to the first and true inventor; and further the inventor must apply for a patent by completing a patent application in accordance with the strict federal patent laws and the rules established by the U.S. Patent and Trademark Office. The patent application acts as both a description of the invention and a request for a patent.

> Note that we said the inventor must "apply" for the patent -- this doesn't mean that the inventor needs to write the patent application or personally correspond with the Patent and Trademark Office. In fact, it's usually not a good idea for the inventor to write the application or carry out the correspondence. (See Section II.2, "Patent Legal Experts," in Part II.)

The Patent and Trademark Office will grant the patent only after they examine and approve the application, and the appropriate fees are paid. In Part V, Chapter 4, "What You Need To Know About Applying For A Patent," we discuss the requirements for a patent application. In Part V, Chapter 6, "What Happens After The Application Is Filed," we explain how a patent application is examined and the routes an application can take before it's approved.

Even though only the inventor may apply for and be granted a patent, the inventor may transfer all or part of the ownership of his patent to another person, and that person becomes the sole or part owner of the patent. (See Part V, Chapter 8, "Patents As Personal Property," which also includes a discussion on when a company doesn't automatically own an engineer's patent.)

III.4 <u>When And Why To Evaluate The Invention For Patentability</u>

Although not required by patent law, we recommend that every invention be evaluated for patentability before preparing and filing a patent application. It's a good idea to perform such an evaluation even though your conclusions won't influence the Patent and Trademark Office since, after a patent application is filed, the Patent and Trademark Office will perform their own evaluation. (See Part V, Chapter 3.)

The advantages to having the invention evaluated for patentability <u>before</u> the application is filed with the Patent and Trademark Office are:

(1)   Having the invention evaluated before you file a patent application will give you a reasonable indication of the patent protection your company can get. No money and time will be wasted in preparing a patent application which has little or no chance of resulting in a patent. In other words, it's better to pay a patent lawyer $500 or so to evaluate patentability early on than to end up wasting $9,000 or more trying to get a patent for an unpatentable invention. (See Part V, Chapter 11, "Patent Costs Itemized.")

(2)   The information gained from a patentability study can be used in the development of an "Information Disclosure Statement" which is a required part of the patent application (see Section V.4.5 in Part V, Chapter 4, "What You Need To Know About Applying For A Patent").

III.5 <u>Prior Art: A Major Factor Of Patentability</u>

The importance of "prior art" cannot be overemphasized -- it plays a major role in determining if an invention is patentable and if you should pursue a patent. To be patentable an invention must be "different" from prior art; and the more closely an invention resembles the prior art the less chance you have of protecting the invention via a patent. In fact, you can't get a patent on inventions which are well described by (i.e. included in) prior art.

An invention can be patentable if it's "different" than prior art even though it's not "better" than the prior art. For example, the invention can be patentable even it's short on inventive merit compared to prior art, and/or it's more expensive to produce than prior art. This means that getting a patent is not, by itself, proof that the invention has commercial value.

As we said earlier, prior art is the **PUBLICLY** known cumulative technical experience and knowledge of every person who has ever lived. This means the potential for discovering prior art for a given invention can be overwhelming, especially for some technologies where a lot of work has been done and made public via patents, products, books, photographs, films, drawings, etc. -- all of which contribute to prior art. However, under U.S. patent law not all prior art affects the patentability of an invention, i.e. not all prior art is used as a "reference" against the invention. Prior art which can be used as "reference" is determined primarily by (a) when, where, and in what form the information became publicly known or available; and (b) when the steps to get a patent were completed. Here's how the law defines prior art which can (and can not) be used as a reference to prevent a patent from issuing -- that is, the prior art that the invention must be compared to, and must be different from, before a U.S. patent will be granted:

(1)   The prior art is a "reference" if it's the same or similar to the invention and was known or used by others in the United States or patented or described in a printed publication anywhere in the world before the date of the invention. (See Part V, Chapter 1 for what the law means by "public use," "printed publication," and "date of invention."

(2)   The prior art is a "reference" if it's the same or similar to the invention and was patented or described in a "printed publication" anywhere in the world more than one year prior to filing the patent application.

> The right to obtain a patent is in fact lost if the party <u>applying</u> for the patent describes the invention in a "printed publication" anywhere in the world more than one year before applying for a patent. (See Part V, Chapter 1.)

(3)   The prior art is a "reference" if it's the same or similar to the invention and in public use or offered for sale by anyone in the United States more than one year prior to the filing of the patent application.

The right to obtain the patent is in fact lost if the party applying for the patent places the invention in public use, offers the invention for sale, or sells the invention in the United States more than one year before filing the patent application (see Part V, Chapter 1).

Figure 1, "The Pyramid of Patentability," illustrates the relationship of prior art to patentability. Figure 1 depicts the following scenario:

(1)   The inventor of the first wheel obtained a patent on the kind of wheel you see in the old cartoon, "The Flintstones;"

(2)   A later inventor obtained a patent on the bicycle wheel; and then,

(3)   Today's automobile wheel was patented.

We will refer to Figure 1 during other discussions in this book, but for now let's use it to examine further prior art as it relates to patentability.

Everything included within the solid line boundaries of Figure 1 is prior art; but note that the three dotted blocks are **NOT** included within the solid lines. These dotted blocks -- being outside of the solid lines -- illustrate a major condition of patentability; i.e. for an invention to be patentable the "technology claimed" in the patent application can not be included in the prior art which was used as a reference against the invention. (See Part V, Chapter 4 for a discussion of "technology claimed.") Also, it's important to realize that the technology claimed by both unexpired and expired patents contributes to prior art and can be used as a reference against future patents. This is depicted in Figure 1 by the continuation of the pyramid under the dotted blocks.

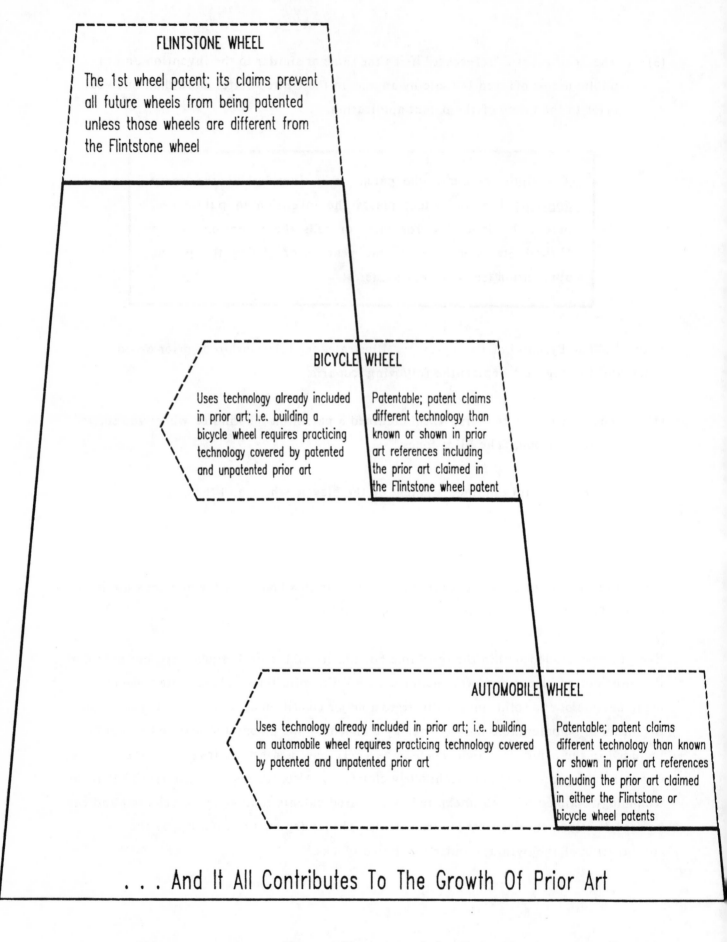

FLINTSTONE WHEEL

The 1st wheel patent; its claims prevent all future wheels from being patented unless those wheels are different from the Flintstone wheel

BICYCLE WHEEL

Uses technology already included in prior art; i.e. building a bicycle wheel requires practicing technology covered by patented and unpatented prior art

Patentable; patent claims different technology than known or shown in prior art references including the prior art claimed in the Flintstone wheel patent

AUTOMOBILE WHEEL

Uses technology already included in prior art; i.e. building an automobile wheel requires practicing technology covered by patented and unpatented prior art

Patentable; patent claims different technology than known or shown in prior art references including the prior art claimed in either the Flintstone or bicycle wheel patents

. . . And It All Contributes To The Growth Of Prior Art

Figure 1:   The Pyramid Of Patentability

Listed below are the types of inventions as they relate to prior art, and how a lawyer should react to them. In every case, a competent lawyer will analyze the invention and prior art using established legal tests and report his conclusions -- all before he starts to prepare a patent application. (See Part V, Chapter 3, "How to Evaluate Obviousness: A Major Factor Of Patentability.")

(1) *Lack of Novelty Inventions (Not patentable)*

Some inventions (but not many) are identical to prior art. In these cases, they are described as "lacking novelty" or as being "anticipated by the prior art" or "old technology." Such inventions can not be protected by patents; and lawyers cannot ethically file, or participate in the filing of, patent applications for these types of inventions.

(2) *Obvious Inventions (Not patentable)*

If the prior art teaches or suggests the invention to a person of ordinary skill in the technology, the invention is "obvious" and therefore cannot be protected by a patent. Competent and ethical patent lawyers will not participate in filing patent applications for inventions that are obvious from knowing the prior art.

(3) *Questionably Patentable Inventions*

It's possible to have an invention which is neither clearly obvious nor clearly unobvious. In these cases, the invention is "questionably patentable" and a good patent lawyer's report will state that "patent protection may be available." Then, if you still want to apply for a patent, the lawyer will prepare an application for that invention. However, you should be cautious when recommending the amount of time and money to be spent on patents for "questionably patentable" inventions. Here's why:

(a) Even if a patent is granted, the patent may be worthless and the validity of the patent may be successfully challenged -- and your company may ultimately lose the patent rights.

(b) Any patent protection obtained will probably have limited commercial value.

17

(c)   Circumstances may change and the original business reason for going after the patent may disappear.

(d)   The cost of obtaining the patent may be far greater than the value of the invention.

(4) Patentable Inventions

Some inventions are clearly and significantly different from prior art. These inventions usually have the greatest and longest lasting commercial value.

III.6 Patent Rights

It's important to understand the legal rights which are granted to the patent owner. An unexpired patent gives the owner the right to <u>exclude others</u> from making, using, or selling the technology claimed in the patent; but <u>does</u> <u>not</u> give the owner the right to manufacture, use, or sell his own invention. The rights that a patent owner has are dependent upon:

(a)   whatever general laws might be applicable, and

(b)   the patent rights of others -- specifically, others may have prior or subsequent superior patent rights covering the invention.

To illustrate this concept of patent rights, let's refer back to Figure 1. Also, let's ignore the fact that patents expire.

(1)   Since the Flintstone wheel patent is the first patent on a wheel, it's called a "basic" or "pioneer" patent; and, because it's the first wheel patent (and assuming it was written to cover all later wheels) the Flintstone patent prevents future wheels from being patented UNLESS they are different from the Flintstone wheel.

(2)   Therefore, the bicycle wheel patent was granted because the bicycle wheel was different from the Flintstone wheel (shown by the dotted block on the right side of the bicycle wheel section). The bicycle wheel patent prevents others from making, using, or selling bicycle wheels which include those differences.

(3)   The automobile wheel patent was granted because the automobile wheel was different (shown by the dotted block on its right side) from BOTH the Flintstone wheel and the bicycle wheel. The automobile wheel patent prevents others from making, using, or selling automobile wheels which include those differences.

(4)   Since the Flintstone wheel patent is the basic patent, the Flintstone wheel patent owner can manufacture, use, or sell his Flintstone wheel as long as he doesn't break any general laws while he's doing it. However, he can't manufacture, use, or sell the bicycle wheel without permission from the bicycle wheel patent owner; and if he wants to manufacture, use, or sell the automobile wheel he needs permission from BOTH the bicycle wheel and automobile wheel patent owners because to build an automobile wheel requires using some of the bicycle wheel's patented differences.

(5)   The bicycle wheel patent owner cannot manufacture, use or sell the Flintstone wheel -- or even his own bicycle wheel -- without permission from the Flintstone wheel patent owner because to build a bicycle wheel or Flintstone wheel requires incorporating some of the technology patented by the Flintstone wheel patent; and if he wants to manufacture, use, or sell the automobile wheel he needs the permission of BOTH the Flintstone and automobile wheel patent owners since the patented differences of the Flintstone wheel must be used in order to build an automobile wheel.

(6)   The automobile wheel patent owner is in even worse shape. He cannot manufacture, use, or sell the Flintstone, bicycle, or even his own automobile wheel without permission from each of the other two patent owners because to build an automobile wheel requires practicing the patented differences of both the Flintstone and bicycle wheels.

(7)   All three patents are known as "mutually blocking" patents.

(8)   All others cannot manufacture, use, or sell:

    *   Flintstone wheels without permission from the Flintstone wheel patent owner, or

19

* Bicycle wheels without permission from the Flintstone and bicycle wheel patent owners, or

* Automobile wheels without permission from the Flintstone, bicycle, and automobile wheel patent owners.

Fortunately, patents do expire after 17 years and inventions can be made, used, and sold without having to go to the stone age owners for permission.

It's interesting to note that as a result of today's advanced state of technology the above scenario does fit some current patent cases. Today, the vast majority of new patents cover improvements or refinements to earlier, more basic inventions; and usually the claims allowed on such improvement patents are limited (or "narrow") as compared with the claims of the more basic patent which can be "broader." This means that the owner can't make, use, or sell his invention without practicing some of the same technology already covered by prior patents; and if he wants to make, use, or sell his invention he must first get permission from the owners of those prior patents. See Part V, Chapter 4 for a discussion of claims.

III.7 Patent Infringement

A patent is infringed when the invention is manufactured, used, or sold -- without permission from the patent owner -- during the time that the invention is protected by a patent. A patent owner has the right to sue in the federal district courts to collect compensation for past infringement and the right to obtain a court order preventing further infringement.

Infringing on another's patent can be expensive. For example, in 1986 a federal court ruled that Eastman Kodak was infringing on some of Polaroid's camera and film patents. As a result, Kodak cannot make, use, or sell its instant cameras or any film for those cameras. Further, Kodak recalled the 16,000,000 infringing cameras, shut down a $200,000,000 plant, and laid off 800 permanent employees and 3700 part-time employees. Beyond that, there are Kodak's 10-plus years of legal fees and the monetary damages (being measured in billions of dollars) which may yet be awarded to Polaroid! We will discuss patent infringement in Part V, Chapter 7, "Avoiding Patent Infringement."

III.8 Maintenance Fees And The Life Of A Patent

Utility and plant patents protect an invention for 17 years from the issue date and then the patent expires. The life of a design patent is only 14 years.

During these years utility patent owners must pay maintenance fees in order to keep their patent in force; and if these fees are not paid on schedule, the utility patent -- and all of the legal rights it provides to the owner -- will expire and the invention will no longer be legally protected. (Note: Maintenance fees are not required for design or plant patents.) To keep a utility patent in force in the United States, the first fee, which is currently $450, must be paid 3-1/2 years from the patent's issue date; the second fee, currently $890, must be paid 7-1/2 years from the issue date; and the last fee, currently $1,340, must be paid 11-1/2 years from the issue date. Independent inventors, small businesses, and non-profit organizations can obtain a 50% reduction in each of these fees.

The fees can be paid within six months after the due date without loss of patent rights. There is, however, a $110 surcharge for a large company (and $55 for a small company or an individual) for paying the fee after the due date but within the six-month period. If the fees are not paid within the six-month period, the patent will expire and the invention will no longer be legally protected.

Under normal circumstances, a patent CAN NOT be renewed. However, certain patents relating primarily to pharmaceutical drugs may be extended for a period of time equal to the time the drug was awaiting approval by the U.S. Food and Drug Administration.

III.9 Patent Rights After The Patent Expires

After a patent expires the technology disclosed and claimed in the patent becomes part of the public domain and anyone may make, use, or sell the invention without the patent owner's permission, provided that in doing so other unexpired patents are not infringed upon. All legal rights associated with a patent, including the right to collect royalties for the use of the patented invention, expire when the patent expires. It's important to realize that the technical information disclosed in an expired patent is prior art; and the existence of this prior art can stop related inventions from being patented. (See Part V, Chapter 3, "How To Evaluate Obviousness: A Major Factor Of Patentability.")

III.10 How To Use Patents As A Valuable Source Of Information

Anyone is free to study a patent once it's issued by the Patent and Trademark Office. This means the issued patents of others are available to you as a valuable source of technical information -- a source that is often overlooked by engineering personnel. You should take advantage of this vast storehouse of technical knowledge, not only for your own research and development efforts, but to help you avoid infringing on another's patent. See Part V, Chapter 10, "How And Where To Find Patent Information," for a discussion on how to get this information.

III.11 Finding Your Way Around The Patent Document

Modern U.S. patents are laid out in the standard format of Figure 2. Even though most of the information in Figure 2 is self-explanatory, the following is a brief explanation of some items. (Note: If information is not available or necessary for a given patent, that particular information field for that patent is left blank.)

United States Patent [19]

Harding et al.

[11] **Patent Number:** **4,584,893**

[45] **Date of Patent:** **Apr. 29, 1986**

[54] **LUBRICATION OF RACK AND PINION APPARATUS**

[75] Inventors: **Peter E. Harding**, West Down, nr. Ilfracombe; **Richard G. Symes**, New Barnstaple, both of England

[73] Assignee: **Harcross Engineering (Barnstaple) Ltd.**, Barnstaple, England

[21] Appl. No.: **474,548**

[22] Filed: **Mar. 11, 1983**

[30] **Foreign Application Priority Data**

Mar. 17, 1982 [GB] United Kingdom 8207827

[51] Int. Cl.⁴ F16H 1/04
[52] U.S. Cl. **74/422**; 74/89.17; 74/467; 184/99; 184/100; 184/5; 308/3.5; 277/153
[58] Field of Search .... 308/3.5, 5 R, DIG. 9; 277/152, 153; 74/89.17, 422, 467; 184/99, 109, 5, 100; 428/408

[56] **References Cited**

U.S. PATENT DOCUMENTS

661,683	11/1900	Ball et al.	308/3.5
990,637	4/1911	Dawson	184/99 X
1,489,580	4/1924	Lucey	74/89.17
1,850,070	3/1931	Booth	74/422
2,361,211	10/1944	Kalischer	308/DIG. 9
2,387,872	10/1943	Bell	308/DIG. 9
2,421,543	6/1947	Cook	308/DIG. 9
2,589,582	3/1952	Strughold et al.	184/99 X
2,919,682	1/1960	Sung	74/422 X
3,012,448	12/1961	Abraham	74/422 X
3,377,799	4/1968	Geyer	74/89.17
3,718,209	2/1973	Moslo	184/100 X
3,745,850	7/1973	Bayle	74/422
3,762,240	10/1973	Adams	74/422 X
3,777,580	12/1973	Brems	74/422 X
3,777,722	12/1973	Lenger	308/5 R
3,841,723	10/1974	Kelso	308/187.1
4,043,620	8/1977	Otto	308/187.2
4,093,578	6/1978	Vasiliev et al.	428/408 X
4,157,045	6/1979	Suzuki	74/467 X
4,280,741	7/1981	Stoll	277/152

FOREIGN PATENT DOCUMENTS

538404	1/1956	Italy	184/99
1219471	1/1971	United Kingdom	74/422
2001409	1/1979	United Kingdom	74/89.17
150329	11/1961	U.S.S.R.	184/99

Primary Examiner—William F. Pate, III
Assistant Examiner—Shirish Desai

[57] **ABSTRACT**

A rack and pinion apparatus has lubrication for the rack (**14**) provided by a split bush (**20**) of graphite or other solid lubricant. The bush is pressed into contact with the surface of the rack by means of a circumferential spring (**22**), a sealing lip (**23**) being provided between the spring (**22**) and the bush (**20**) to seal the interior of the apparatus against dust and other foreign matter. The graphite lubricates not only the bearing surfaces (**15** and **16**) for the rack (**14**) but also the teeth on the rack and the pinion (**11**).

6 Claims, 1 Drawing Figure

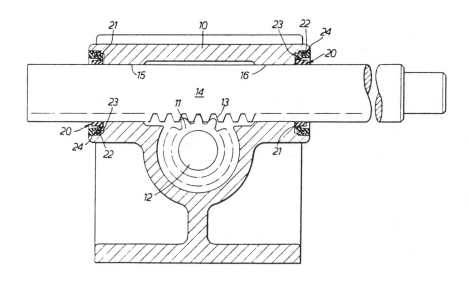

Figure 2: Picture Of A Patent

(Page 1 of 2)

1

LUBRICATION OF RACK AND PINION APPARATUS

This invention is concerned with the lubrication of rack and pinion apparatus.

Die lifters are used extensively in the car industry and to meet the demands of the production line must operate reliably for many thousands of cycles with minimal maintenance. Die lifters are essentially rack and pinion devices and satisfactory lubrication is a problem that has existed for many years. In particular, it is now the practice in the car industry to steam clean dies and this practice very effectively cleans grease lubricants out of the die lifters requiring them to be re-lubricated.

It is therefore an object of the present invention to provide improved lubrication of rack and pinion apparatus.

The present invention is rack and pinion apparatus comprising a rotatable pinion, a reciprocating rack having teeth in engagement with the teeth of the pinion, bearings for said rack and a supply of lubricant for said bearings, said supply comprising solid lubricant constrained to press against said rack while allowing the rack to move relative to the lubricant.

The lubricant may be provided in the form of a split bush.

Preferably a constraining spring is provided around the periphery of the bush to press the internal bush surface onto the rack.

A sealing lip may be provided between the spring and the periphery of the bush.

An embodiment of the present invention will now be described, by way of example, with reference to the accompanying drawing which is a cross-section of a rack and pinion gearbox.

Referring now to the drawing, a rack and pinion gearbox is provided with a housing **10** in which is mounted a pinion **11** on a shaft **12**. The shaft **12** is supported in nylon bushes (not illustrated) filled with a solid lubricant, namely molybdenum disulphide. Teeth **13** on a reciprocable rack **14** engage the teeth of the pinion, and the rack **14** is mounted on bearing surfaces **15** and **16** in the housing **10**.

At the outer end of each of the bearing surfaces **15** and **16** is provided a supply of lubricant in the form of a split bush **20** of solid lubricant, in this embodiment graphite. The bush **20** has an external flange **21** located against the housing **10**. A circumferential constraining spring **22** bears on a sealing lip **23** to seal it against the periphery of the bush **20** and so retains the halves of the bush **20** in contact with the shaft despite wear. The sealing lip **23** is preferably elastomeric and is bonded to a metal reinforcement ring **24** which is mounted in and seals against the housing **10**. In this way the bushing is retained in the housing by the flange between the sealing lip and the housing and foreign matter such as dust is excluded from the interior of the housing.

2

In use, the lubricant from the bush **20** is deposited on the surface of the rack **14** and conveyed onto the bearing surfaces **15** and **16**. Also particles of lubricant are shaved off the bushing by the teeth of the rack and are deposited on the pinion thus providing satisfactory lubrication for the gearbox over a long period. It should also be noted that if the bush **20** wears out it is a quick and simple operation to replace it with a new bush.

It has also been found, somewhat surprisingly, that the bushing **20** and the lubricant coating on the various surfaces survive steam cleaning so that the apparatus requires no lubrication.

Molybdenum disulphide is also a suitable solid lubricant.

We claim:

1. A solid lubricating arrangement for rack and pinion apparatus which converts an input rotary motion into an output linear motion or vice versa, comprising:
   a housing,
   a pinion supported for rotation in said housing,
   a rack supported for reciprocating movement in said housing, said rack having teeth in engagement with teeth of the pinion,
   said housing forming bearing surfaces interiorly of said housing for supporting said rack and resisting substantially all transverse forces to which said rack is subjected from the outside,
   a supply of lubricant for each of said bearing surfaces, each supply comprising a solid lubricant material located next to and separate from a side of each of said bearing surfaces along the direction of said rack toward the exterior of said housing, and
   means fixed to said housing in the vicinity of said bearing surfaces for urging each said supply of lubricant to press against said rack including the teeth thereof while allowing the rack to move relative to the lubricant material, so that the lubricant material is carried by said rack to be deposited on said bearing surfaces when said rack reciprocates in said housing.

2. Apparatus as claimed in claim 1, in which the solid lubricant material is provided in the form of a split bush seated in said housing.

3. Apparatus as claimed in claim 2, in which a constraining spring is provided around the periphery of the bush to press the internal bush surface onto the rack.

4. Apparatus as claimed in claim 2, in which an annular flange is provided on the outer periphery of the bush to confront said housing.

5. Apparatus as claimed in claim 3, in which an annular sealing lip fixed to said housing is provided to extend radially between the spring and the flange on the periphery of the bush for retaining the bush in said housing and preventing matter other than said solid lubricant material from entering the interior of said housing.

6. Apparatus as claimed in claim 1, wherein said solid lubricant material is selected and arranged to be carried by the teeth of said rack for deposition on the teeth of said pinion.

* * * * *

65

Figure 2: Picture Of A Patent

(Page 2 of 2)

(1)   *Patent Number*: The patent number, which is always printed in the upper right-hand corner of the patent, is the primary method of identifying patents. You need only the patent number to obtain a copy of the patent from the Patent and Trademark Office. In fact, trying to locate a patent by any other identification (e.g. XYZ Company's patent on a particular subject) would probably require the services of a professional patent searcher. (For a discussion on professional searchers, see Part V, Chapter 10.)

(2)   *Issue Date*: This date, printed directly beneath the patent number (shown as "Date of Patent" in the example), is the date on which the patent was issued by the Patent and Trademark Office. This patent, being a utility patent, expires 17 years from this date.

(3)   *Title:* The title is a short, descriptive term which generally identifies the subject matter of the patent. Usually, the title is vague (shown in our example as "Lubrication of Rack and Pinion Apparatus") and is of little or no value in identifying a patent.

(4)   *Inventors*: Here, the inventor or inventors are identified by name and address.

(5)   *Assignee*: This identifies the <u>owner</u> of the patent (not the inventor) as shown in the Patent and Trademark Office records. The absence of a name here does not necessarily mean that the patent has not been assigned by the inventor, but could mean that the Patent and Trademark Office was not requested to show an assignee on the published patent.

(6)   *Appl. No.*: This is the application number that the Patent and Trademark Office assigned to the patent application. The numbers range from 1 to 999,999 and are repeated in a new 1 to 999,999 series as necessary. This means that over a period of time there can be two patents with the same <u>application</u> (not patent) number.

(7)   *Filing Date*: The filing date (shown as "Filed" in the example) is the date on which the complete patent application was received by, or in certain cases mailed to, the Patent and Trademark Office. (See Section V.4.9 of Part V, Chapter 4.) The filing date is important because:

(a)   It determines which patents, technical developments, and other information (i.e. prior art) can be used as a reference against the application. (See Section III.5, "Prior Art: A Major Factor Of Patentability.")

(b)   It is used for settling disputes over who was the first inventor of the same invention (i.e. interferences) and therefore who is entitled to the patent. See Part V, Chapter 1, "How And Why To Record The Important Dates," for a discussion on interferences.

(c)   It plays a major role in patent infringement situations.

(8)   *Related U.S. Application Data*: This field lists any U.S. patents or applications by the same inventor which are closely related to the patent itself. There is no Related U.S. Application Data field on our sample patent because there is no related U.S. application data for this particular patent.

(9)   *Foreign Application Priority Data*: This field contains information identifying the foreign counterpart application, if any, on which the inventor is claiming his "right of priority." (See Part V, Chapter 9, "Foreign Patents.") Our sample patent states that the original patent application on this invention was filed on March 17, 1982 in Great Britain as British application number 8207827.

(10)   *Int. Cl.*: This is an international classification number which identifies the technical category and subcategory in which the patent is classified according to an international patent classification scheme. This system is used by most countries of the world and provides a means for locating foreign patents which relate to the same general subject matter.

(11)   *U.S. Cl.*: This U.S. classification field identifies the technical category and subcategory into which the Patent and Trademark Office has classified the patent, allowing for a convenient way to find most other U.S. patents relating to similar inventions. The details of this classification system are available in the publication, *Manual of Patent Office Classification*, and its corresponding publication, the *Index to the Manual of Patent Office Classification*. Both of these are available from:

Superintendent of Documents
U.S. Government Printing Office
Washington, DC 20402

and are also available in major libraries. In addition, almost all patent lawyers keep up-to-date copies of both publications.

(12) *Field of Search*: This identifies the U.S. classification categories that the Patent and Trademark Office examiner searched during his examination of the application.

(13) *References Cited*: These are a listing of prior U.S. and foreign patents or publications which the patent examiner felt were the most pertinent to the patented invention.

(14) *Primary Examiner* and *Assistant Examiner*: These fields identify the examiners in the Patent and Trademark Office who examined the application and determined that the invention was patentable.

(15) *Abstract*: This is a brief description of the invention. It allows the reader to decide quickly if he should spend more time studying that particular patent.

(16) *Drawing*: The drawing (or drawings) shows the details of one or more structures which include and describe, or "embody," the invention. The drawings are intended only to show the structures containing the invention and do not limit the invention in any way. Consequently, it's possible that a device is covered by the patent even though the device doesn't look like the drawing in the patent.

(17) *Main Text*: The main text of the patent (shown on the second page of Figure 2) provides information about the invention, its background, and applications. This section is typically broken into the following components which may or may not be identified by subheadings:

   (a) *Background of the Invention*: A general description of what the inventor recognizes and acknowledges to be the prior art. Often, it also includes the inventor's interpretation of the problem(s) or disadvantage(s) associated with the prior art.

(b) *Summary of the Invention*: A brief explanation of what the inventor believes his invention or contribution to the art is. Here the inventor may describe the advantages his invention has over prior art.

(c) *Brief Description of the Drawings*: A short explanation of each of the various views of the drawings.

(d) *Detailed Description of the Drawings*: This is a specific item-by-item description of all parts shown in the drawings describing the structural and functional interrelationship of the parts which make up the invention, and the operation of the invention as intended by the inventor. This discussion often goes into great detail specifying materials, sizes, structural features, etc. Patent law states that (1) the invention must be described in such clear, concise, and exact terms that any person skilled in the technology would be able to understand, make, and use the invention by following the description; and (2) the information in the description must identify the best manner known to the inventor for practicing the invention.

(e) *Claims*: These usually brief, one-sentence statements are what patents are all about. That is, they:

1. identify the unique features of the invention which make the invention different from other prior inventions (and therefore make an invention patentable),

2. define the legal rights protected by the patent, and

3. are the basis for determining patent infringement.

We will discuss the claims and their importance in greater detail in Part V, Chapter 4, "What You Need To Know About Applying For A Patent."

III.12 When And Why You Need A Patent Lawyer or Agent

Unless you are skilled in patent law and experienced in working with the Patent and Trademark Office, it is best to have a patent lawyer or agent help with most of your patent activities.

Engineers can and should do some things on their own, for example, engineers can and should:

*   Keep records of the dates and events surrounding inventions (Part V, Chapter 1), and

*   Perform the simpler patent searches (Part V, Chapter 10);

However, the primarily legal, non-technical patent activities are generally complex, time consuming, and usually better and more cost-effectively performed by a patent lawyer or agent. These are:

*   Carrying out the difficult patent searches (Part V, Chapter 10),

*   Evaluating an invention for obviousness (Part V, Chapter 3),

*   Preparing a patent application and writing claims (Part V, Chapter 4), and

*   Prosecuting the application before the Patent and Trademark Office (Part V, Chapter 6).

Further, if you perform these activities on your own, you may get a patent but if you haven't done it right, the patent may (a) not properly cover the invention, or (b) be challenged later and declared invalid. In either case, the invention will not be properly protected and most probably your company will have wasted the effort and money put into obtaining the patent.

PART IV. A COMPANY PATENT POLICY TO FOLLOW

A company should have a patent policy. Engineering personnel must be aware of the significance of, and know how and when to document, the important dates discussed in Part V, Chapter 1. Further -- and even if the company has an in-house patent lawyer -- each of the company's operating units should designate a person for collecting and reviewing the unit's inventions. Inventions and the status of those inventions should be monitored to maintain the company's right to a patent and to maximize the patent protection which can be obtained.

When the operating unit first determines that an invention is important enough to be patented, an invention disclosure should be prepared and immediately afterwards a completed copy of the "Invention Disclosure Data Form" (provided in Part V, Chapter 2) should be sent to the patent lawyer. In turn, the patent lawyer should conduct a patentability study, which may include an evaluation for obviousness as summarized in Part V, Chapter 3. The patent lawyer should then issue a report which elaborates on prior technology and the potential for patent protection. Then, the engineering department and perhaps other management personnel should decide if a patent should be pursued.

A company's patent policy should state that the company will:

I.   Have agreements with all engineering, technical, and management employees relating to the ownership of inventions, patents, copyrights, and trade secrets made during the performance of their jobs. (See Section V.8.4, "When *Doesn't* The Company Own The Engineer's Patent?" in Part V, Chapter 8, "Patents As Personal Property," and Parts VI and VIII for discussions on copyrights and trade secrets.)

II.   Encourage commercially valuable inventions.

III.   Invest in legal and technical effort to insure the highest quality patents and protection, and obtain and maintain all patent protection to which they are legally entitled.

IV.   Insure that patent costs are not needlessly incurred.

PART V. PATENT DETAILS ENGINEERING PERSONNEL NEED TO KNOW

It's not necessary that you know as much about patents as does a patent lawyer but there are some details you must know about patents. Here's a summary of those details and where to find them in this book.

CHAPTER 1:   **How And Why To Record The Important Dates**
In this chapter we tell you how to properly record the important dates; and if you don't, why your company may lose their rights to get a patent.

CHAPTER 2:   **How And When To Report An Invention To Your Patent Lawyer**
You must provide the patent lawyer with the information he needs about your important inventions. For this purpose, we include and discuss an "Invention Disclosure Data Form."

CHAPTER 3:   **How To Evaluate Obviousness: A Major Factor Of Patentability**
This chapter outlines the method used by the Patent and Trademark Office, patent lawyers, and the courts to determine whether an invention is patentable from the very important standpoint of obviousness.

CHAPTER 4:   **What You Need To Know About Applying For A Patent**
This chapter gives the background details engineering personnel should know about who may apply for the patent, what the application must include, and the like.

CHAPTER 5:   **Reviewing The Patent Application**
Since the law holds the <u>inventor</u> responsible for the accuracy of a patent application, in this chapter we include a detailed checklist and discussion to allow <u>you</u> to easily check if the application is complete and accurate.

CHAPTER 6:   **What Happens After The Application Is Filed**
Here, we explain how the patent application is treated by the Patent and Trademark Office from the moment the application is received, including what can be done to reverse negative decisions. You will know what to expect (and when to expect it) after filing a patent application.

CHAPTER 7:   **Avoiding Patent Infringement**
This chapter tells what you, as an engineering person, need to know about
patent infringement: what it is, what the penalties for infringement are,
how patent infringement can be avoided, and the proper use of patent
symbols. This information will help you guard against the serious situation
of infringing on the patents of others.

CHAPTER 8:   **Patents As Personal Property**
In this chapter we discuss the options available to the patent owner; e.g.
how his patent or patent application can be licensed, sold, or transferred to
another. This discussion will help you with such dealings (and deals) with
other companies and individuals. We also explain when a company does
and doesn't own inventions made by engineering personnel.

CHAPTER 9:   **Foreign Patents**
This information will help you contribute to your company's filing for
foreign patents. We explain the value of foreign patents, foreign patent
laws as they differ from U.S. patent laws, what the differences mean to
engineering personnel, and what is required before one can apply for
foreign patents.

CHAPTER 10:   **How And Where to Find Patent Information**
The information contained in the patents of others can be very valuable to
an engineering department. To this end, this chapter tells how and where
to find every U.S. patent and how to keep up to date on the newest patents.
We also examine the information you might want from patents and how to
go about getting it.

CHAPTER 11:   **Patent Costs Itemized**
This is a summary of what it costs to search existing and expired patents;
and to apply for, receive, and maintain a patent. It will help you
recommend prudent patent activities for your inventions.

PART V. CHAPTER 1: **HOW AND WHY TO RECORD THE IMPORTANT DATES**

The exact dates on which you take certain actions when developing and using an invention -- and being able to prove those dates -- can be vital because:

(1)   Your company's right to obtain a patent <u>can</u> be lost if good records of the invention are not maintained.

(2)   Your company's right to obtain a patent <u>will</u> be lost if the patent application is not filed within certain time periods.

(3)   You <u>may</u> become involved in a conflict if another company files an application at the same time for substantially the same invention and you may have to prove the dates before you can obtain a patent.

V.1.1 <u>Six Important Dates</u>

Make sure you keep good records of the following six dates:

(1)   The date the invention is conceived,
(2)   The date the invention is reduced to practice,
(3)   The date the invention is first used in "public,"
(4)   The date the invention is first "published,"
(5)   The date the invention is first offered for sale, and
(6)   The date the invention is first sold.

A discussion of these important dates follows.

V.1.2 <u>The First Two Dates (And Diligence) Fight Interference</u>

Sometimes two (or even more) inventors file a patent application at about the same time for substantially the same invention. Since in the United States a patent can only be granted to one inventor, i.e. the first inventor, the Patent and Trademark Office carries out a proceeding known as an "interference" in order to determine who is the first inventor and therefore is entitled to the patent. If you have no other appropriate evidence to submit to the Patent and Trademark Office, you will be restricted to using the filing date of the patent application as the earliest date of your invention (see Part V, Chapter 6, "What Happens After The Application Is Filed"). Consequently, for reasons of potential interference, always keep records of the following two dates.

(1)   *The Date of Conception*: This is the date the invention was first "conceived" and one of the two important dates in deciding who wins in an interference proceeding. In patent law "conception" is:

> "the formation of an idea complete or concrete enough to enable one of ordinary skill in the pertinent art to proceed toward the completion of the invention without the exercise of inventive skill."

This means that an invention has been "conceived" when it's to the point where an average, knowledgeable engineering person could take it over and complete it without making additional inventions.

Since conception is generally a mental act, the fact that an idea was actually conceived can usually be proven only by some form of objective, tangible evidence; e.g. a written description, sketch, drawing, or model. Therefore, for all practical purposes the date of conception becomes the date the written description, sketch, drawing, or model was <u>attested</u> <u>to</u> <u>by</u> <u>witnesses</u>.

The inventor who can prove that he <u>conceived</u> the idea first has one of the essential elements necessary to win an interference.

(2)   *The Date of Reduction To Practice*: "Reduction to practice" is the act of completing an invention after having conceived of the invention, as described above. Patent law recognizes two types of reduction to practice -- constructive and actual.

"Constructive reduction to practice" doesn't require the construction or use of the invention, but only requires that a <u>patent application</u> be filed which describes the invention in complete enough terms to: (a) show that the invention will work, and

(b) allow another person having ordinary skill in the art to make and use the invention.

"Actual reduction to practice" means that a working model of a device which includes the invention has been constructed and operated satisfactorily. An invention is <u>not</u> reduced to practice if the model was built but didn't work. It's important to know that building a working model of the invention <u>is</u> <u>not</u> a substitute for filing a patent application. A patent will not be issued unless a patent application is filed with the Patent and Trademark Office.

The Date Of Invention: The date that the invention is reduced to practice by the same person who conceived the invention.

Interference occurs in only about 1% of the patent cases, but it might as well happen 100% of the time if it happens to you. Interferences are resolved based on:

(a)   who has the earliest dates for conceiving the invention and reducing the invention to practice; and

(b)   how "diligent" the inventor was in reducing his invention to practice after the date of conception.

Legally, diligence requires a "substantially unbroken effort" from the date of conception to reduction to practice. This means that if Inventor A conceived an invention before Inventor B, but Inventor B reduced the invention to practice before Inventor A, whether Inventor A is the first inventor will depend on how "diligent" Inventor A was in reducing his invention to practice. If your company has not been diligent in reducing one of your inventions to practice, in an interference your company will not have the best case for proving that you are the first inventor -- even though your company may have the earlier date of conception.

The law also says that you're NOT the inventor if you reduce the invention to practice but you didn't conceive the idea in the first place. You can always have others assist you in

37

constructing a model or otherwise help reduce the concept to practice, but to be the inventor you must have conceived of the invention. This also means that if the person who conceives the idea is no longer with the company, the ex-employee is still the inventor even if another employee of that company reduces the idea to practice.

Interference can exist between an <u>issued patent</u> and a <u>patent application</u> provided the patent has not been issued for more than one year prior to the filing of the application, and provided that the application is not barred from being patentable for some other reason. This means it's possible to lose your patent rights even if someone applies for a patent <u>after</u> your patent is issued. On the other hand, if you're the first inventor and are able to prove the important dates it's possible for you to get a patent even if someone already has a patent on the same invention.

In cases of interference, the decision on who is entitled to the patent is made by the Board of Patent Interferences which is part of the Patent and Trademark Office. After that, the losing party may appeal to the Court of Appeals for the Federal Circuit or file a civil action against the winning party in the appropriate U.S. federal district court, neither of which is part of the Patent and Trademark Office.

A Quick Summary Of Interference

After you conceive the idea you can perfect your company's patent rights either by (a) building a working model of the invention, or (b) filing a patent application which completely describes the construction and operation of the invention. In any event you must file an application before you can obtain a patent. It's possible that, after you file the patent application, your company can wind up in an interference over who made the invention first. In these cases your company will be legally entitled to a patent on the invention if you can prove by good written records that:

(1)   you have the earliest dates of conception and reduction to practice; or if you don't have the earliest date of reduction to practice,

(2)   you have the earliest date of conception and were diligent in pursuing a reduction to practice.

V.1.3 <u>About Witnesses</u>

Signatures of witnesses are needed on many of the patent records (e.g. drawings, sketches, and descriptions) because in many proceedings in the Patent and Trademark Office (as well as in the courts) <u>unsupported,</u> self-serving testimony is suspect and not admissible as evidence. Hence, the testimony of the inventor alone will not be sufficient to prove the dates of his invention even though his testimony is supported by records which he signed. However, if the inventor's testimony is supported by the testimony of witnesses, the inventor's testimony will be allowed and be more persuasive. The best witnesses are those (a) to whom the invention was explained at or near the time the invention was conceived; (b) who are technically capable of understanding the invention; and (c) who signed and dated the inventor's drawings and descriptions at the time the invention was conceived. Co-inventors cannot act as witnesses since co-inventors' statements about the dates or other activities relating to the invention are also self-serving and therefore not credible as evidence.

V.I.4 <u>The Importance Of The Other Four Dates</u>

(1)   *The Date of First Public Use*: The law states very clearly that your company will lose their right to obtain a U.S. patent unless they file a patent application within one year after the invention is placed in "public use" in the United States. Unfortunately, however, sometimes it's not easy to know when a "use" is "public use."

        A use **IS** clearly "public use" if the invention is used for commercial purposes. For example, your company has one year to file a patent application after they finalize the invention and use the invention to make money.

        On the other hand, it's **NOT** "public use" if your company uses the invention in front of a person who is in a confidential relationship with your company -- <u>provided</u> that the invention is not being used for commercial purposes. A confidential relationship is one in which the parties agree to maintain information in confidence and not to disclose such information to others. (See Part VII, "Secrecy Agreements.") Accordingly, in these cases your company's rights to obtain a patent are not affected.

You can usually consider any person employed by your company, or having a common interest with your company, as already having a confidential relationship with your company. However -- and contrary to popular belief -- suppliers normally do not have a common interest with customers, and to be safe you should create a specific written confidential relationship with a company that you may be considering as a source of supply for your invention.

When it comes to the right to get a patent, of particular interest to engineering personnel is the testing of an invention; i.e. when does "experimental use" become "public use?" Here are some benchmarks:

(a)   Laboratory testing -- The one-year clock doesn't start to run (i.e. the right to obtain a patent is not affected) if you're testing an invention in the lab and the only persons involved are those who either work for your company or are in a written confidential relationship with your company.

(b)   Field testing -- The one-year clock doesn't start to run if you can prove that the field test is (1) required to prove that the invention works, (2) performed before persons in a confidential relationship with your company, (3) commensurate with the tests required to prove that the invention works properly, and (4) not being done for commercial gain. Usually, all of the above can be proven by producing a document which states why and how the tests are being performed, what role outsiders will play in the testing, keeping appropriate testing records, and the inventor taking an active role in the test.

However, the one-year clock **WILL** start to run if the testing fails to comply with each of the requirements outlined above. For example, recently a farm equipment manufacturer tested its invention by having farmers use equipment which included the invention in the harvesting of the farmers' crops; and then the farmers sold the crops harvested by the farm equipment. Unfortunately for the manufacturer -- and even though the farmers did not know that the invention existed in the equipment -- the farmers' selling of the harvested crops was considered to be commercial gain which caused the "experimental use" to be a "public use" and ultimately prevented the farm equipment manufacturer from obtaining a patent.

It's always a good idea to get your patent lawyer's advice on field tests of patentable inventions so that the intended "experimental use" can't be later construed as "public use."

(2)   *The Date of First Publication*: U.S. patent law says:

(a)   Your company cannot get a U.S. patent if your invention was described in a "printed publication" anywhere in the world before the date of your invention (i.e. the date the invention is reduced to practice by the same person who conceived it); and

(b)   Your company will lose their right to obtain a U.S. patent if it does not apply for a patent within one year after the invention is disclosed in a "printed publication" anywhere in the world.

It's very important to realize what the law defines as a "printed publication." The law considers the work to be "printed" if it's handwritten, typed, or printed; and considers the work to be a "publication" when it becomes available to the public -- even if its availability is only one copy in an obscure public library.

(3)   *The Date of First Offer for Sale*: The right to obtain a U.S. patent is lost if your company does not apply for a patent within one year after the invention, or a device which includes the invention, is offered for sale in the U.S.

This area of patent law is not always black and white. For example, if your company builds a device which includes the invention and tries to sell it, they clearly have made an offer for sale under the patent laws. However, if they have not built a device but still try to interest a customer in buying one, they may or may not have made an "offer for sale" depending upon a variety of surrounding facts. Sometimes it's tough to decide if, in the eyes of the law, an "offer for sale" was made and it's best to contact your patent lawyer for an opinion if a questionable situation arises.

(4)   *The Date of First Sale*: The right to obtain a patent is lost if the application for a patent is not made within one year after the invention, or device which includes the invention, has been sold in the U.S. ("Sold" does not necessarily mean that money changed hands; e.g. the invention may be traded.) This rule is absolutely black and white. If your company does not file for a patent within one year after their first sale they won't get a patent -- it's that simple.

V.1.5   How To Develop Patent Sketches and Descriptions

To properly document an invention you should make numbered and dated sketches of the invention, and numbered and dated written descriptions of the sketches.

The sketches should show all of the significant details of the invention. To facilitate description and aid in understanding, each significant part on the sketch should be given a different number.

The description should support the sketch by describing how the invention works and what is new about it. It should explain each significant part shown on the sketches and emphasize how the invention differs from the prior art. Be sure to include the first descriptions that were made of the invention and any available laboratory and engineering reports.

Above all, the sketch and description must completely yet simply describe the idea. There are two reasons for this:

(1)   For all practical purposes, the date of conception will be the date the written description and sketches (or a model) were attested to by witnesses, and these witnesses must fully understand the idea without any additional explanation.

(2)   It should be assumed that the patent lawyer knows nothing about the technology or problems concerning the invention. Although patent lawyers are graduate

42

engineers it's virtually impossible for them to know about all technological fields, especially when it comes to specialized research and engineering.

V.1.6 How To Document The Dates

There are no right or wrong ways to record the important dates but here are some suggestions. Documenting the date of conception and diligence to reduce the invention to practice requires more attention than documenting any of the other dates.

You should take at least the following steps when documenting the date of conception:

(1)   When you get an idea or concept for a potentially patentable invention -- as soon as possible -- develop a sketch of the idea and write a brief description of it, preferably in ink. The combination of the sketch and description should explain the idea well enough so that someone of ordinary skill in the technology can fully understand the idea without any additional explanations.

(2)   Sign and date both the sketch and description. If there are two or more inventors, each should sign both the sketch and description. Use the date on which the sketch and description were actually signed -- not the date the idea was thought of.

(3)   As soon as possible, show the sketch and written description to at least two co-workers or other trustworthy persons who can act as witnesses. As we said earlier, the date that the sketch and written description are witnessed by others becomes the date of conception. Be certain the witnesses fully understand the idea behind the invention. The witnesses should sign and date the sketch and written description.

> In the next chapter you will find an "Invention Disclosure Data Form." Part I of that form can be used to make a record of the evidence described in Items (2) and (3) above.

(4)   Promptly store the original signed and witnessed document in a safe location.

When the sketch and description are completed and witnessed, you have not only recorded the "date of conception" of the invention but you also have created an important legal document. This document will make it possible for you and the witnesses to testify accurately, even years later if necessary, that the invention was conceived on that day.

IMPORTANT

If, in the past, you made an invention but did not make an original sketch or drawing and written description of the invention or did not have it signed, dated, and witnessed <u>DO</u> <u>IT</u> <u>NOW</u>. Use the date of <u>actual</u> signing on the sketch and description; that is, <u>do</u> <u>not</u> backdate the sketch, your signature, or the signatures of the witnesses. Such backdating would result in falsified evidence which is worse than no evidence at all.

Diligence to reduce the invention to practice can be documented by maintaining project progress via engineering notebooks, project reports, and the like. These records should document efforts and events of significance and summarize the work performed on your project. You should sign and date each entry and periodically (at least once each month) have the record reviewed and witnessed by two other technically competent persons. These records can then be used if it becomes necessary to prove diligence in the reduction to practice of the invention.

The other dates (i.e. reduction to practice, public use or publication, offer for sale, and actual sale) can often be proven from engineering project or test records, purchase orders, work orders, and other similar records created during the normal course of business. However, if you are in a rapidly evolving field with substantial patent activity or if your records do not allow you to reasonably establish these dates, a separate project log <u>should</u> <u>be</u> <u>maintained</u>. The log may be an engineering notebook and should include regular periodic entries concerning progress of the project, test results, changes, etc. The entries should preferably be made in ink, signed, witnessed, and dated. These entries do not have to be witnessed by the same people who witnessed the date of conception.

V.1.7 <u>Summary</u>

Being able to prove the six important dates on which you or your company take certain actions in developing and using an invention, and your diligent effort to reduce your invention to practice after you conceive it, can be crucial to your company obtaining a valid patent. The recording of two of these dates (the date of conception and the date of reduction to practice) as well as the recording of your effort to reduce the invention to practice, is especially important if another inventor files an application at about the same time for substantially the same invention; or as a reference for interference with recently issued patents. In each of these cases, the Patent and Trademark Office carries out an interference proceeding to determine who is the first inventor and therefore is entitled to the patent. The other four dates (first public use, first publication, first offer for sale, and first sale) are of special importance with regard to on-time filing of the patent application. If a patent application is not filed within one year of any of these four acts your company will lose the right to obtain a patent. If you keep a signed and dated written record of your invention activities, and have all drawings and descriptions witnessed by people who understand the invention, you can be reasonably sure that not only will the patent application be filed on time but that you will be prepared in case of interference proceedings.

PART V. CHAPTER 2:   HOW AND WHEN TO REPORT AN INVENTION
                      TO YOUR PATENT LAWYER

Engineering departments should adhere to the policy described in Part IV, "A Company
Patent Policy To Follow." This policy includes (a) keeping records of all inventions
according to Section V.1.6, "How To Document The Dates," in Part V, Chapter 1, and (b) for
all of the important inventions, promptly completing the following "Invention Disclosure
Data Form" and sending a copy of the form to the patent lawyer. Timely and proper
completion of an invention disclosure form which discloses the details of your invention
will allow the patent lawyer to preserve your company's legal rights to your important
inventions.

IMPORTANT

The instructions for completing the "Invention Disclosure Data Form"
also provide additional information you need to know about patents.

After the form is completed, keep the original in a safe, permanent file and send a copy to
your company's patent lawyer. You are free to copy the "Invention Disclosure Data Form"
included here for use as your own invention disclosure data form.

Figure 3: Invention Disclosure Data Form

PART I: Sketches, Drawings, and Descriptions

Complete and attach a copy of this sheet to every sketch, drawing, and description of those sketches and drawings. Be sure to number each of your sketches, drawings, and descriptions. Also be sure that the individuals who sign as witnesses fully understand your invention and that they provide their full post office addresses so they can be located later if necessary. This sheet along with your sketches, drawings, and descriptions then become good evidence for documenting the date of conception of your invention.

Reference number assigned to the sketch, drawing, or description _____
Number of pages _____
Date of the sketch, drawing, or description _____

Inventor #1:   **Inventor #2:**
Name (Print)_____   Name (Print)_____
Address_____   Address_____

_____   _____
Citizenship_____   Citizenship_____
Date_____   Date_____
Signature_____   Signature_____

Witnessed and Understood By: **Witnessed and Understood By:**
Name (Print)_____   Name (Print)_____
Address_____   Address_____

_____   _____
Date_____   Date_____
Signature_____   Signature_____

NOTE: Do not show the attached information to any person not employed by your company. Doing so may result in a loss of patent and trade secrets rights.

Figure 3: Page 1 of 6

48

PART II: Pertinent Dates and General Background

Please print or type the following information.

Case No. _____ (to be assigned by patent lawyer)

INVENTION TITLE _____ DIVISION _____

1.   The date the invention was first conceived (thought of): _____

2.   Names and addresses of others able to understand the invention to whom the invention was first disclosed and the date of those disclosures:

_____

_____

_____

_____

3.   The date that the first sketch or drawing was made and where that sketch or drawing is presently located: _____

4.   The names of the individuals to whom the first sketch or drawing was disclosed and the date of that disclosure:

_____

_____

_____

5.   The date of the first written description and where that description is presently located:_____

6.   The names of the individuals to whom the first written description was disclosed and the date of that disclosure:

_____

_____

_____

Figure 3: Page 2 of 6

7.   Was a working prototype constructed? Yes ___ No ___
      If yes:
      a. Date started: _____
      b. Date completed: _____

8.   Was the working prototype successfully operated?
      Yes ___ No ___ If yes, state place and nature of first successful operation:
      _____
      _____

9.   Names of witnesses to the first successful operation of the invention and the dates
      they witnessed that operation:
      _____
      _____
      _____

10.   To the best of your knowledge, was the invention known or used by others, in public
      use, offered for sale, or sold in the United States; or patented or described in a
      "printed publication" anywhere in the world? See Part V, Chapter 1 for a definition
      of the terms used in this question. Yes ___ No ___
      If yes, give any dates and circumstances you know of:
      _____
      _____
      _____
      _____

11.   Is the invention in production? Yes ___ No ___
      a. If yes, date of first production: _____
      b. If no, date when production is contemplated _____

Figure 3: Page 3 of 6

12.   Proposed use(s) of invention:

_____

_____

_____

13.   Project(s) name and/or number(s): _____

_____

14.   Was the invention conceived while working under a government contract?

Yes ___ No ___

If yes:

a. Contract No.: _____

b. Classification: _____

15.   Were construction and operation done under government contract?

Yes ___ No ___

If yes:

a. Contract No.: _____

b. Classification: _____

Figure 3: Page 4 of 6

PART III. Invention Disclosure

(Use additional sheets if necessary)

1.   Briefly describe the problem or need which led to the invention.

2.   Describe or list prior art known to you, including identifying patents or publications of which you are now aware. Be sure to discuss the advantages of your invention over these prior art devices.

3.   List those features of your invention which you think may be novel.

Figure 3: Page 5 of 6

52

4.   Sign your full name(s) (including your unabbreviated full first name and middle name), your full post office address, and your country of citizenship. Also have two witnesses who understand the invention sign and date this form in the spaces allocated. Be sure to have the witnesses provide their full post office addresses so they can be located later if necessary.

Inventor #1:

Name (Print)_____

Address_____

_____

Citizenship_____

Date_____

Signature_____

Inventor #2:

Name (Print)_____

Address_____

_____

Citizenship_____

Date_____

Signature_____

Witnessed and Understood By:

Name (Print)_____

Address_____

_____

Date_____

Signature_____

Witnessed and Understood By:

Name (Print)_____

Address_____

_____

Date_____

Signature_____

Figure 3: Page 6 of 6

V.2.1 Explanation of the Invention Disclosure Data Form

The following explains how to accurately fill out the Invention Disclosure Data Form of Figure 3. If this form is properly completed, you will provide your patent lawyer with what he needs to know about the invention. You are free to copy Figure 3 for this purpose.

Completing Part I: Sketches, Drawings, and Descriptions

Complete and attach a copy of page 1 of the form to each of the first sketches or drawings (or copies of these first sketches or drawings) of the invention. Note that page 1 requires assigning numbers, numbers of pages, and dates to the sketches and drawings. Follow this same procedure for descriptions of the inventions; i.e. assign numbers and complete and attach a copy of page 1 to each description. In addition, the inventor -- and the two witnesses to the sketches, drawings, and descriptions -- should sign and give their full names and addresses. To avoid self-serving -- and therefore non-permissible -- evidence it's important that the two witnesses review and understand the invention and that no co-inventor sign as a witness for any document.

Completing Part II: Pertinent Dates and General Background

Items 1 through 10 (Pages 2 and 3) -- Your company's U.S. patent rights may be lost if another company beats you to the punch in conceiving an invention and reducing it to practice. Further, a patent cannot be obtained if the application is not filed within one year after the invention has been placed in "public use," offered for sale, sold, or described in a printed publication. (See Part V, Chapter 1.) Items 1 through 10 are designed to let your patent lawyer know of the activities surrounding these potentially damaging situations.

The witnesses chosen for Items 2, 4, 6, and 9 should be people who are capable of understanding the invention. Usually these are other engineers, draftsmen, technicians, or (in rare cases) even members of the inventor's family.

Item 11 (Page 3) -- This informs the lawyer of the production status of the invention. This information will assist him in determining when he should start a patent infringement investigation and when to file a patent application in U.S. and foreign countries.

Item 12 (Page 4) -- Here, describe for what application(s) the invention is intended.

Item 13 (Page 4) -- Here, supply the name(s) and/or number(s) of the project(s) under which the invention was conceived and/or developed.

Items 14 and 15 (Page 4) -- In these sections, reveal whether the invention was conceived, constructed, and/or operated while working under a government contract; and whether the invention is classified as secret, top secret, etc. This has significance in that many U.S. government contracts require the inventor to: (a) report any inventions made under or in the course of the contract of the government; and (b) give the government title to or a license to use the invention.

Completing Part III: Invention Disclosure

Item 1 (Page 5) -- This is self-explanatory in the form.

Item 2 (Page 5) -- Here is where to inform the patent lawyer of known, pertinent prior art. Supplying this information will help the lawyer understand the principal advantages of the invention before he advises on filing a patent application; and, if the decision is to file, these facts are essential to the application's Information Disclosure Statement (see Section V.4.5 in Part V, Chapter 4) and to the writing of the all-important patent claims.

Prior art (which we introduced in Section III.5, "Prior Art: A Major Factor Of Patentability," in Part III) is what has already been done and is known in the related technology. Prior art includes:

(1) The prior inventions of others (patented, unpatented, or even if the patent expired);

(2) Prior commercially available devices;

(3) Prior publications; and

(4) Any other prior information which is relevant to the patentability of the invention.

> **IMPORTANT**
>
> The patent applicant -- and anyone who is substantively involved in preparing the patent application -- has a duty to disclose any pertinent prior art he knows of. This means that the patent lawyer must reveal what he knows about relevant prior art; and depending upon their involvement, the obligation may also extend to the engineering manager and even co-workers. In all cases these disclosures must be made in absolute candor and good faith.
>
> Although no one is required to conduct research to find the most pertinent prior art, the duty includes disclosing to the Patent and Trademark Office all known prior art which would reasonably be considered pertinent to the invention, and the duty applies even where the disclosure of prior art may result in a refusal of the patent. Failure to make this disclosure may be a "fraud on the Patent and Trademark Office" which could invalidate any patent obtained on the invention, and possibly have other bad legal consequences for the inventor and the company.

Almost always, engineering personnel have special knowledge about prior art which they learned from technical literature, patents, closely related products, etc. In these cases, the lawyer needs the corresponding magazine citations, patent numbers, name of the manufacturer, etc. The prior secret work of co-workers -- even though the work is related to the invention for which the application is being filed -- does not have to be disclosed to the Patent and Trademark Office, but the patent lawyer should be made aware of such work so that he can decide if it's prior art.

In addition to revealing what the inventor and others know about prior art, in Item 2 the inventor should also include an explanation of the advantages of his invention over the prior art and of any shortcomings of such prior art. These shortcomings may be either technical or economical.

Item 3 (Page 5) -- Here, explain the aspects of the invention which are novel; i.e. what is thought of as being new and different from the prior art.

Item 4 (Page 6) -- In this section, the inventor should print or type his full name, address, and country of citizenship. This information will ultimately be required by the Patent and Trademark Office. In addition, two witnesses should also give their complete names and addresses. (This will assist in locating them at a later date.) These two witnesses, who need not be the same two witnesses who signed page 1 of the form, must also have reviewed the invention and understood it.

The inventor(s) and the witnesses should sign the "Invention Disclosure Data Form" in the spaces provided for signatures, and enter the date on which they signed. Note that there are spaces provided for the signatures and addresses of a joint inventor, and you can add more if necessary. Remember, no co-inventor should sign as a witness for any document.

V.2.2 Final Tip: Don't Delay

It's important that a patent application be filed as soon as possible so that your competitors do not have an opportunity to file their patent applications before you do. If some of the information requested by the "Invention Disclosure Data Form" is unavailable or difficult to get, fill it out as completely as possible and send it to your patent lawyer. Never delay the patent process for any length of time just because some of the information is unavailable.

V.2.3 Summary

If you complete the "Invention Disclosure Data Form" for every one of your important inventions -- and do so on a timely basis -- you will be giving your patent lawyer the information he needs when he needs it. In turn, the lawyer can take the steps necessary to preserve all of your company's proprietary patent rights.

PART V. CHAPTER 3:   **HOW TO EVALUATE OBVIOUSNESS: A**
                       MAJOR FACTOR OF PATENTABILITY

Very few inventions are based on a completely new technology; i.e. there are prior technical publications, patents, and sometimes even products which also discuss or include the same or similar technology. Therefore, when testing an invention for patentability, it's generally necessary to evaluate the invention for obviousness.

V.3.1 Three Conditions For Patentability

Figure 4, "Is It Patentable And/Or A Trade Secret?," illustrates when an invention is patentable and how to handle the invention if it's not patentable. Note that even if the invention is not patentable it still may have value as a trade secret (see Part VIII, "Trade Secrets") and, failing that, the engineering effort most probably contributes to your base of engineering concepts and experience.

As shown across the top of Figure 4, there are three conditions which must be met before an invention is patentable:

(1)   The invention must be useful. Only in rare cases does this condition inhibit the inventor from getting a patent; especially in your environment where your company -- and every other company in the world -- pays engineers to work on only useful devices.

(2)   The invention must be novel; i.e. it can not be identical to the prior art.

(3)   The invention taken as a whole must not be obvious to a person having ordinary skill in the art. In a nutshell, this condition of "obviousness" relates to how closely the invention resembles the prior art and what constitutes "ordinary skill" in the technology. (See Section III.5 in Part III for a discussion on prior art and an introduction to obviousness.) This chapter elaborates on obviousness.

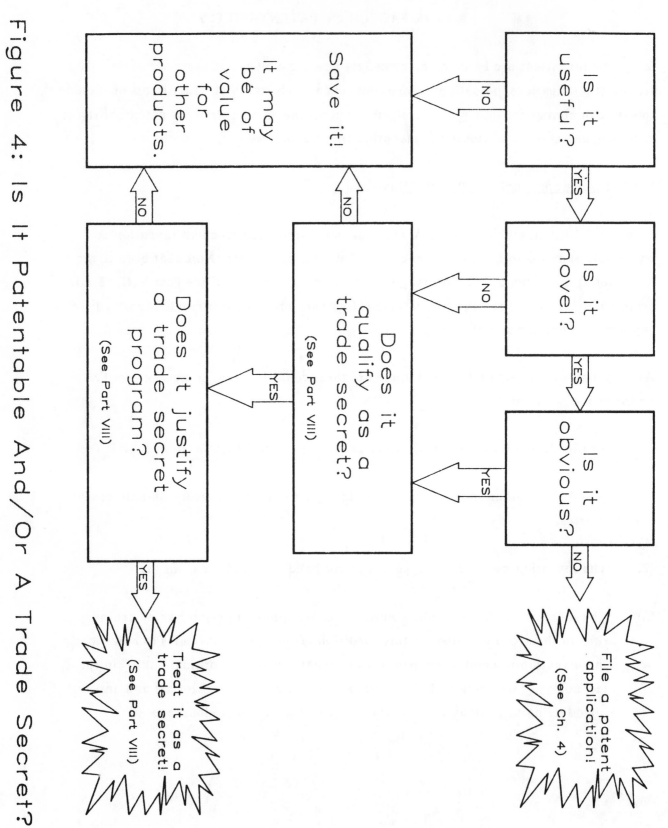

Figure 4: Is It Patentable And/Or A Trade Secret?

V.3.2. Procedure For Evaluating Obviousness

In the test for patentability, determining obviousness is the most difficult activity because it often involves subjective judgments and conclusions. However, the law provides a short test for obviousness by asking the question:

> "At the time the invention was made, were the differences between the invention taken as a whole and the prior technical information obvious to someone with ordinary skill in the pertinent technology?"

If the answer to this question is clearly yes, the invention is obvious and therefore not patentable. Or, to put it another way, if the prior art TEACHES or SUGGESTS the invention to a person with average knowledge of the technology, then the invention is "obvious" and therefore unpatentable. For the statute regarding obviousness, see S103 in Appendix D.

Although there is no succinct, clear way of evaluating the obviousness of an invention, the following is a summary of the four-step procedure used by patent lawyers, the Patent and Trademark Office, and the courts to reach a conclusion regarding obviousness.

Step 1: Identifying the Prior Art -- Each invention should be compared with the most pertinent prior art, including obvious extensions of prior art. Usually, the most pertinent prior art can be found by searching through applicable patents located in the Patent and Trademark Office's public search room in Arlington, Virginia. Each pertinent patent is carefully analyzed and described in writing. Often, one single patent (the most pertinent patent) can be identified.

You or your patent lawyer may perform the search or you may hire a searching specialist to do it for you. (Patent searches are discussed in Part V, Chapter 10, "How And Where To Find Patent Information.")

Step 2: Ascertaining the Differences Between the Invention and the Prior Art -- The differences between the invention and the most pertinent prior patent(s) are then determined. The differences, which usually become apparent when making an in-depth analysis of the patent(s), are identified and listed.

61

Step 3: Resolving the Level of Ordinary Skill Required in the Pertinent Art -- The level of skill, knowledge, and experience possessed by the <u>average person</u> skilled in the pertinent art must be established and becomes one of the standards upon which obviousness is decided. Making this determination is usually not easy and requires judgments but can be estimated by answering the following questions.

*   What level of education is common?

*   How much experience is common for engineers and technicians working in the industry?

*   What level of skill is shown in other pertinent patents?

Step 4: Evaluating the Secondary Considerations of Obviousness -- In addition to carrying out Steps 1, 2, and 3 there are some secondary factors which must be examined before an obviousness conclusion can be made. Some of these are: commercial success, technical bias (i.e. a commonly shared technical opinion that the approach is wrong), long-felt need, failure of others to produce the invention, and copying.

Evaluating these secondary considerations can get complicated. For the purposes of this book it's sufficient to say that the <u>results</u> of these factors must correlate directly with the proposed invention or they are not even considered in reaching a conclusion on obviousness. For example, let's assume that the patent application is for an invention which is already included in a product being sold to the public by the same company applying for the patent. The fact that the product which includes the invention is selling well does not render the invention unobvious if the reason for the sales is not related to the invention. That is, maybe the invention is very obvious to anyone having ordinary skill in the art but heavy advertising, and not the invention itself, produced the large sales. Likewise, long-felt need, technical bias, failure of others to produce the invention, and the fact that someone else copied the invention may or may not be persuasive evidence of obviousness (either positively or negatively) depending upon the surrounding circumstances.

Upon evaluating the invention and comparing it with prior art, a good patent lawyer should be able to say with a relatively high degree of confidence whether the invention is sufficiently different from prior art to be patentable or too similar to the prior art to be patentable. Usually, if the patent lawyer finds the invention to be obvious, you can then correct any misunderstandings or supply the lawyer with additional facts which may change his opinion. If, after again reviewing the facts, the lawyer concludes that the invention is patentable, you can then ask him to prepare a patent application.

WARNING

Regardless of what conclusion the patent lawyer makes about the obviousness of the invention, it's usually left up to the engineering department (or perhaps higher management based on the advice of engineering personnel) to decide whether to proceed with the patent application. However, keep in mind that if you decide to file an application against the lawyer's advice the examiner from the Patent and Trademark Office, others who are interested in your patent, and the courts (if the patent is ever contested in court*) will also conduct an evaluation for obviousness and apply the same rules which the patent lawyer used. Consequently, there is a probability that if the patent lawyer finds the invention to be obvious, so will others in the chain. There's a chance you may be right in pursuing a patent against the advice of a patent lawyer but you stand a better chance of being wrong.

V.3.3 <u>Summary</u>

There are three conditions which must be met before an invention can be patented: the invention must be useful, the invention must be novel, and the invention must not be obvious. Most often, the invention will be an improvement on prior art inventions; that is, there are patents or other published information which deal with the same or similar technology as does the invention for which a patent is being sought. Therefore, the Patent and Trademark Office always examines the patent application to make sure that the invention and the prior art are sufficiently different so that the prior art does not teach or

* *Ultimately, only a court can decide whether or not an invention is obvious.*

63

suggest the invention to someone with average skill in the technology. If the prior art and the invention are so closely related that the invention is "obvious" to someone having this average skill, the Patent and Trademark Office will not grant the patent. To avoid wasting time, money, and effort in pursuing a patent your company's inventions should be evaluated by your patent lawyer for patentability <u>before</u> the patent application is filed. Such an evaluation will also assist in the development of an "Information Disclosure Statement" which is needed for the patent application.

PART V. CHAPTER 4:   **WHAT YOU NEED TO KNOW ABOUT**
                     APPLYING FOR A PATENT

Before a patent is granted, the inventor must first file an application with the Commissioner of Patents and Trademarks in the Patent and Trademark Office. The patent application acts as both a description of the invention and a request for a patent.

V.4.1 _Who May Apply For A Patent_

Patent law dictates that _only_ the _inventor_ may apply for a patent; and, if a person who is not the inventor applies for and is granted a patent, the patent could be invalid.

> Although we say that the inventor must "apply" for the patent, this doesn't mean that the inventor needs to write the patent application himself. Patent applications are complex legal documents which must comply with detailed requirements and, even though the law allows for an inventor to prepare and submit a patent application on his invention, the help of an experienced patent lawyer or agent is almost always needed.

The patent application must be filed in the inventor's name even if the patent will subsequently be assigned to the company or if the inventor dies before assigning the patent to the company. If the inventor dies before the assignment to the company, his administrator or heir (e.g. his wife) then becomes the inventor's representative and the company must deal with that person for completing the application and assignment of the invention to the company. If the inventor becomes unavailable or incompetent before executing the assignment, or has a clear obligation to assign patents to the company but refuses to do so, a representative acting on behalf of the inventor may file for the application, but the application would still have to be in the inventor's name. Regardless of the situation, the reasons for persons other than the inventor being involved in the application must be explained to the Patent and Trademark Office.

People who work together on the technical aspects of an invention are called joint inventors and must <u>jointly</u> apply for a patent. Sometimes deciding who is and who isn't a joint inventor can be difficult and in these cases it's best to ask your patent lawyer for an opinion. A clear-cut example of joint inventorship is where several engineers working together have a concept for an invention which includes contributions from each. However, in cases where a person only makes a financial contribution to the invention rather than a technical contribution, that person is not a joint inventor and should not be named in the application as an inventor.

An application for a patent must be accompanied by an oath or declaration of inventorship. If one signs such an oath knowing he is not a true inventor, he would be committing perjury and could be prosecuted and subjected to a fine or imprisonment. An innocent mistake of failing to name a joint inventor or naming the wrong person as an inventor can generally be corrected.

In the United States any inventor, regardless of his citizenship, may apply for a patent -- and the same rules apply to everyone whether they are U.S. citizens or not. This means that a foreign applicant can also submit his own U.S. patent application, but if he chooses to have a representative represent him he too must use a patent lawyer or agent who is registered to practice before the U.S. Patent and Trademark Office.

V.4.2 <u>What The Application Must Include</u>

A patent application <u>must</u> be accompanied by the following:

A.   A specification including both a description of the invention and the claims which are being made.

B.   An oath or declaration of inventorship.

C.   A drawing of the invention (when required).

D.   The appropriate filing fee.

Also, applicants are encouraged to file an Information Disclosure Statement, i.e. a statement of the prior art known to the inventor and others substantively involved in preparing the application.

These requirements are discussed below.

V.4.3 The Document

The document -- which must be legibly written, printed, or typed in English, in permanent ink on only one side of the paper -- must contain the specification and oath or declaration of inventorship.

A. **Specification Portion of the Document**

The specification makes a full disclosure of the invention which, when the patent is issued, will be available to the public. This public disclosure is the basis for providing inventors with the protection afforded by the patent laws. (See Section III.1, "Why Do Patents Exist?" in Part III.)

The specification must teach those of ordinary skill in the pertinent art how to practice the invention or the patent will not be issued by the Patent and Trademark Office; or, if the patent is issued, the patent will be invalid. The specification must describe the invention fully, clearly, concisely, and precisely and describe how the invention differs from pertinent prior art, including prior expired and unexpired patents. It must describe completely the process, machine, manufacture, composition of matter, or improvements invented, and the invention's principle mode of operation (if applicable). The inventor must provide what he believes is the best mode of carrying out the invention. This requirement prevents inventors from keeping to themselves the best way of carrying out the invention. For example, let's say you're developing an electronics device and, even though you know that the best semiconductor to use in the device is one from the XYZ Company, you call out a less desirable semiconductor from the ABC Company. In this case, the patent could be invalidated.

When the invention is an improvement on another invention the specification should not describe the prior art in painful detail but should be confined to -- and elaborate on -- the details of the improvement.

The specification need not include any commercial advantages (such as lower cost or ease of manufacture) but if such commercial advantages are discussed they must be as accurate as possible and not be misleading.

The specification should include the following information:

First, on the first page a short and specific title of the invention should be shown as a heading.

As an alternative, a more elaborate introduction can be used which states not only the title but also the citizenship and residence of the applicant.

Second, the abstract of the disclosure.

The abstract of the disclosure is a brief (four or five sentence) description of the invention which will ultimately appear on the published patent, allowing someone to get a quick and concise idea of what the invention is. The abstract is usually typed on a separate sheet as the last page of the application.

Third, a cross-reference to any related pending patent applications.

This is more complicated than it sounds since there are several types of related patent applications. "Continuations," "continuations in part," "divisionals," and "streamlined divisionals" are common terms used by the Patent and Trademark Office to identify patent applications which have something in common (e.g. the inventor, the invention, etc.) and are pending in the Patent and Trademark Office.

Fourth, a brief summary of the invention.

The summary is written to allow the public to determine what the invention is. It should indicate the invention's nature, substance, and objective; and usually discusses prior art. Whereas the abstract of the disclosure (explained above) is limited to only a few sentences, the summary has no such limitation.

Fifth, a brief description of all the views of any drawings.

Drawings must be provided when they will help a reader to understand the invention. This means that practically all inventions (except some that relate to a chemical or metallurgical composition) require drawings.

Sixth, a detailed description of the invention.

This part of the specification must fully describe the invention, including how to make and use the invention. It must be written in clear, concise, and exact terms so that any person skilled in the technology, or the technology with which the invention is most nearly connected, can make and use the invention. This description must also refer to the different views and parts of the drawing.

Lastly, the claim or claims.

The specification must end with one or more claims defining the invention; that is, brief descriptions of the invention that clearly identify the invention's unique features including those features which distinguish it from earlier inventions. THE CLAIMS ARE ALL-IMPORTANT -- they are the measure of the protection which is obtained in a patent. Initially, the Patent and Trademark Office bases its decision to grant a patent on the claims, and after the patent is granted the claims are used to judge questions of validity and infringement.

Claims <u>must</u> be written in a single sentence.

When more than one claim is presented, some of the claims may refer to prior claims. Claims which stand alone, i.e. have no reference to another claim, are called "independent claims." Claims which refer to prior claims are called "dependent claims" and are used to narrow the claim on some portion of the invention that was defined by the claim it refers to.

This "narrowing" is best explained by example. In Figure 2, "Picture Of A Patent," in Part III:

Claim 1 (the independent claim) paraphrased says, "We claim a solid lubricating arrangement . . . comprising a solid lubricating material . . . "

Claim 2 (the dependent claim) says, "Apparatus as claimed in Claim 1, in which the solid lubricating material is provided in the form of a split bush seating in said housing."

Claim 1 is "broader" than Claim 2 since Claim 1 says that anyone using a solid lubricating material -- any solid lubricating material -- would be infringing. Claim 2 is "narrower" than Claim 1, and is said to "more completely cover the invention" than Claim 1, since Claim 2 says that infringement would take place -- not because someone is using a solid lubricating material -- but because they are specifically using "a solid lubricating material which is in the form of a split bush housing seated in a housing."

In addition, a patent application can contain multiple dependent claims; i.e. claims that narrow more than one claim. A multiple dependent claim must refer to prior claims in the alternative. This means a multiple dependent claim must say "a device as described in claim A **OR** claim B," it can't say "a device as described in claim A **AND** claim B."

All of the claims must use words or phrases that were used in the description of the invention. That is, the terms and phrases used in the claims, and the claims themselves, must clearly conform to, and be supported by, the description of the invention so that any confusion caused by the claims can

70

be cleared up by referring to the description. The Patent and Trademark Office also requires appropriate fees to accompany the claims (see Section V.4.6, "The Filing Fee," in this chapter and Part V, Chapter 11, "Patent Costs Itemized").

Because of their importance, the writing of the claims requires the services of a trained patent lawyer or a skilled and experienced person who is very familiar with the prior art, patent law, and Patent and Trademark Office procedural requirements.

B. Oath or Declaration of Inventorship Portion of the Document

The Patent and Trademark Office requires the inventor to make an oath or declaration of inventorship. It must be in English unless the inventor does not understand English, in which case the oath can be in a language the inventor understands together with its English translation. An oath or declaration consists of the inventor signing a statement which identifies the specification to which it's directed and states the following:

(1)   That the inventor has read and understands the contents of the specification and claims.

(2)   That the inventor believes himself to be the original, first, and sole (or joint) inventor of the invention claimed in the application.

(3)   That the inventor acknowledges the duty to disclose to the Patent and Trademark Office information which is material to the examination of the application.

(4)   That the inventor is claiming "right of priority" benefits based on a foreign patent application if such an application exists; and, if so, the oath should include information about the foreign application. (See Part V, Chapter 9.)

The inventor should sign the oath or declaration of inventorship after he reviews the patent application and agrees that it is accurate and complete (see Part V, Chapter 5, "Reviewing The Patent Application"). Patent law states that the inventor should sign the oath or declaration immediately after reviewing the patent application but not more than six weeks before the application is filed in the Patent and Trademark Office. The Patent and Trademark Office presumes that an oath or declaration which is more than six weeks old is stale. Therefore, if the application is not filed within six weeks after the oath or declaration is signed, the inventor should review the application again and sign a new oath or declaration of inventorship. The oath or declaration must be sworn before a Notary Public or other officer authorized to administer oaths. The signing of an oath or declaration containing willful false statements is punishable by fine, imprisonment, or both.

If a patent lawyer represents an inventor in corresponding with and prosecuting an application in the Patent and Trademark Office, the inventor must first give the lawyer a power of attorney to do so. Since such representation is common practice, patent lawyers combine the oath or declaration, power of attorney, and other basic requirements into a single document approved by the Patent and Trademark Office. Figure 5, "Declaration And Power Of Attorney For Patent Application," is an example of such a form. You're free to copy Figure 5 for this purpose.

Figure 5: Declaration And Power Of Attorney For Patent Application

As a below named inventor, I hereby declare that:

My residence, post office address and citizenship are as stated below next to my name.

I believe I am the original, first and sole inventor (if only one name is listed below) or an original, first and joint inventor (if plural names are listed below) of the subject matter which is claimed and for which a patent is sought on the invention entitled

_____,

the specification of which

(check one)   ___ is attached hereto

              ___ was filed on _____ as

              Application Serial No. _____

              and was amended on _____ (if applicable).

I hereby state that I have reviewed and understand the contents of the above specified specification, including the claims, as amended by any amendment referred to above.

I acknowledge the duty to disclose information which is material to the examination of this application in accordance with Title 37, Code of Federal Regulations, S1.56(a).

I hereby claim foreign priority benefits under Title 35, United States Code, S119 of any foreign application(s) for patent or inventor's certificate listed below and have also identified below any foreign application for patent or inventor's certificate having a filing date before that of the application on which priority is claimed.

Prior Foreign Application(s)   Priority Claimed

_____ _____ _____   ___ Yes ___ No

(Number)   (Country)   (Day/Month/Year Filed)

_____ _____ _____   ___ Yes ___ No

(Number)   (Country)   (Day/Month/Year Filed)

I hereby claim the benefit under Title 35, Unites States Code, S120 of any United States applications(s) listed below and, insofar as the subject matter of each of the claims of this application is not disclosed in the prior United States application in the manner provided by the first paragraph of Title 35, United States Code, S112, I acknowledge the duty to disclose material information as defined in Title 37, Code of Federal Regulations, S1.56(a) which occurred between the filing date of the prior application and the national or PCT international filing date of this application.

_____   _____   _____
(Application Serial No.)   (Filing Date)   (Status - patented, pending,

_____   _____   _____
(Application Serial No.)   (Filing Date)   (Status - patented, pending,
                                                            abandoned)

POWER OF ATTORNEY: As a named inventor, I hereby appoint the following attorneys:

_____

full powers of substitution and revocation, and; _____

_____

to prosecute this application and transact all business in the United States Patent and Trademark Office connected therewith.

SEND CORRESPONDENCE TO: _____

DIRECT TELEPHONE CALLS TO: _____

I hereby declare that all statements made herein of may own knowledge are true and that all statements made on information and belief are believed to be true; and further that these statements were made with the knowledge that willful false statements and the like so made are punishable by fine or imprisonment, or both, under Section 1001 of Title 18 of the United States Code and that such willful false statements may jeopardize the validity of the application or any patent issued thereon.

1)   Full name of sole or first inventor _____

     Inventor's signature _____ Date _____

     Residence _____ Citizenship _____

     Post Office Address _____

2)   Full name of second joint inventor, if any _____

     Inventor's signature _____ Date _____

     Residence _____ Citizenship _____

     Post Office Address _____

(Supply similar information and signature for third and subsequent joint inventors)

V.4.4 The Drawing

Drawings are required in most patent applications except where the invention relates to a chemical or metallurgical composition -- but not just any drawing will do. Since (a) all drawings are published in a uniform style when the patent is issued, and (b) they must be readily understood by persons reading the patent, the drawings must be done in a particular form and must show every feature of the invention specified in the claims. The individual parts of the drawings must be referenced by letters or numerals; and the drawings cannot contain names or other identification within their boundaries. The Patent and Trademark Office specifies the size of the sheet on which the drawing is made, the type of paper, the margins, and other details. The complete drawing requirements are specified in *The Rules of Practice in Patent Cases* which is available from:

> Rules Service Co.
> 7658 Standish Place, Suite 106
> Rockville, MD 20855
> (301) 424-9402

V.4.5 The Information Disclosure Statement

The patent application should be accompanied by a statement identifying and discussing pertinent prior art which is known to the inventor, his patent lawyer, and others who are substantively involved in preparing the application. That is, the inventor and the others must reveal what they know about previous work, products, literature, etc. which is prior art and which pertains to the invention in question. There is no legal requirement to conduct a search to determine whether there is pertinent prior art, but if the inventor or others who are substantively involved in preparing the patent application are aware of such prior art they must disclose it to the Patent and Trademark Office. If it's proven that pertinent prior art was knowingly not revealed, the patent, if issued, can be invalid. The prior art statement must include what is known about the invention from:

(1)   The prior inventions of others. The prior secret work of co-workers need not be disclosed to the Patent and Trademark Office but your patent lawyer should be made aware of such work;

(2)   Prior commercially available devices;

(3)   Prior publications;

(4)   Other prior information which is relevant to the patentability of the invention.

See Section III.5 in Part III and Part V, Chapter 2 for additional information on prior art. Contact your patent lawyer if you have questions about these prior art requirements.

V.4.6 The Filing Fee

The patent application will not be processed unless it's accompanied by the proper filing fee. The filing fee is normally paid by check or money order, payable to the Commissioner of Patents and Trademarks. The Patent and Trademark Office assumes no responsibility for its safe arrival.

The basic filing fee is $340 for a utility patent application and $140 for a design patent application. (See Section III.2, "What Can Be Patented?" in Part III for the definitions of utility and design patents.) In addition, there are fees of:

(a) $34 for each independent claim in excess of three,

(b) $12 for each independent and dependent claim in excess of twenty, and

(c) $110 for each application containing one or more multiple dependent claims.

As with maintenance fees (see Section III.8 in Part III) individual inventors, small companies, and non-profit organizations can obtain a 50% reduction in each of the above filing fees.

If after the initial filing of the patent application the number of claims are increased, a fee is due based on this same fee schedule. (Just in case you're wondering, you won't get your money back if you reduce the number of claims!)

Normally there are no additional governmental fees required during the prosecution of the application in the Patent and Trademark Office. However, additional fees can be required during the prosecution stage if due dates are not met when responding to the various "Official Actions" (i.e. the Patent and Trademark Office's responses to a patent application -- see Part V, Chapter 6, "What Happens After The Application Is Filed.")

V.4.7 <u>When Models, Exhibits, and Specimens Are Needed</u>

Years ago, models were required for all inventions which could be modeled. Today, it's just the opposite. The Patent and Trademark Office will not accept a model unless they specifically request one. Instead, they require that the drawings and descriptions of the invention be sufficiently clear and complete to adequately disclose the invention without the need of a model. The Patent and Trademark Office will request a model or other physical exhibit if they believe that one is necessary to allow them to better examine the application. For example, in a recent patent application filed for an alleged perpetual motion device, the Patent and Trademark Office required a working model and had that model examined by engineers and scientists at the National Bureau of Standards.

If the invention involves a composition of matter (i.e chemical compounds, formulas, and the like) the inventor may have to supply specimens of the composition, or specimens of the composition's ingredients or intermediate compounds.

V.4.8 <u>Finishing Up The Application</u>

The patent application must be signed by the inventor or by the person who can lawfully apply for the patent on the inventor's behalf; that is, if the inventor is unavailable, becomes incompetent, is deceased, or refuses to sign the application, a personal representative or assignee of the application may be empowered to sign on behalf of the inventor. (See Section V.4.1, "Who May Apply For A Patent" in this chapter.) Patent law requires that the full name (including the unabbreviated full first name or middle name), the citizenship, and the full post office address be given for each inventor.

When the application is ready for filing and the necessary papers have been signed by the inventor, the application is delivered (usually mailed) with the appropriate fee to:

> Commissioner of Patents and Trademarks
> U.S. Patent and Trademark Office
> Crystal Plaza
> 2021 Jefferson Davis Highway
> Arlington, Virginia 20231

V.4.9 How To Get The Earliest Filing Date

Ordinarily, the filing date given the application is the date that a complete application is received by the Patent and Trademark Office. (See Section V.6.1, "If The Application Is Complete," in Part V, Chapter 6.) However, the filing date will be the date that the application is mailed to the Patent and Trademark Office if (a) the application is sent by the "Express Mail Post Office To Addressee" service offered by the U.S. Postal Service; (b) all papers and fees are identified by the number of the "Express Mail" mailing label; and (c) all papers and fees include a certificate of mailing which states the date of mailing and are signed by the person making the mailing.

V.4.10 Summary

Before a patent can be granted on any invention, a patent application must be filed -- in the inventor's name -- with the Commissioner of Patents and Trademarks in the U.S. Patent and Trademark Office. The application must be filed in the inventor's name even if the patent is, or will ultimately be, assigned to the company he works for; or if the inventor is unavailable for any reason. Due to the complexity involved, it's advisable to get a patent lawyer's help in writing the application. A patent application must contain: a specification and claims; an oath or declaration of inventorship; a drawing, if required; and the proper fee. It's important that the claims be done right since they will be used to judge patentability, validity, and infringement. An application should also be accompanied by an Information Disclosure Statement which reveals all that the inventor, the patent lawyer, and others participating in the preparation of the application know about the pertinent prior art, i.e. the previous work, products, literature, etc. of others. When the application is completed, it should be sent to the Patent and Trademark Office, where it will be numbered, given a filing date, and put on the stack to await its turn for review.

PART V. CHAPTER 5: REVIEWING THE PATENT APPLICATION

Even though it's usually best that a patent lawyer or agent write the patent application, it's essential (and a legal requirement) that each inventor thoroughly review the application because:

(1)   The inventor is responsible for making sure that the information contained in the application is complete and accurate.

(2)   Much of the information required in a patent application is known only to the inventor.

(3)   The inventor is required to sign a sworn statement that he has read, understood, and agrees with the adequacy and correctness of the application; and the inventor can truthfully sign this statement only after he has carefully reviewed the proposed patent application. Signing a false statement can subject the inventor to criminal fines and imprisonment and make the patent obtained on the application invalid.

The checklist on the next page will allow an inventor to systematically examine a patent application to make sure that it completely and accurately describes the invention.

Note: If the inventor is not available, e.g. deceased, insane, incompetent, or otherwise unavailable, the engineering manager may have to assume the responsibility for reviewing the application prior to filing.

<u>Figure 6: Patent Application Checklist</u>

Patent law requires that you, the inventor, sign a sworn statement, i.e. an oath or declaration of inventorship, that among other things you have read and understood the patent application (see Section V.4.3 in Part V, Chapter 4). To this end, review the patent application using this checklist and inform your patent lawyer if you find that the application requires changes or additions.

The patent application must contain:

A.   A specification including a description of the invention and claims to the invention.

B.   An oath or declaration of inventorship.

C.   A drawing of the invention, unless the application relates to an invention where a drawing is not possible.

D.   The proper filing fee.

The documents that you must sign are the oath or declaration of inventorship, the power of attorney (see Figure 5, "Declaration and Power of Attorney for Patent Application" in Part V, Chapter 4) and, depending on your legal obligations to your employer, a document which assigns the invention to your employer. However, <u>do</u> <u>not</u> sign anything until all documents are clear, complete, and accurate.

The filing date of a patent application can have important, beneficial legal consequences; hence, it's important to review your application as soon as possible so that the application will receive the earliest possible filing date by the Patent and Trademark Office.

(To be filled in by patent lawyer.)

Information Disclosure Statement

(1)   Does the information disclosure statement reveal the most pertinent prior art known to you or anyone else substantively involved in the patent application? If not, elaborate on any pertinent prior art which you know of but is not discussed in the application. Prior art includes not only the patents found through the lawyer's searches, but also any prior patents, publications, technical papers, products, etc. which you may know of.

Patent law requires that you, as the inventor, and anyone else who is substantively involved in preparing the application reveal to the Patent and Trademark Office all of the prior art you or they know of. This means that the patent lawyer must reveal what he knows about prior art; and depending upon their involvement, the obligation may also extend to the engineering manager and even co-workers. In all cases these disclosures must be made in absolute candor and good faith.

(2)   Is the discussion about the prior art fair, accurate, and complete? If no, explain.

Specification and Drawings

After the application is filed, very few corrections and practically no additions to the specification and drawings can be made. Therefore, check the application's specifications and drawings carefully.

(1)   Are the specification and drawings clear, complete, and technically accurate? If no, point out the errors.

(2)   Is the invention explained completely and accurately enough to allow a person of ordinary skill in the art to make and use the invention without experimentation or trial and error? If no, explain the additional information or drawings needed.

(3)   Do the specification and drawings disclose the best way to make and use the invention? If no, identify and explain the best way. Patent law says that you must spell out the best way you know of to make and use the invention and if you don't the patent (if granted) could be invalid.

(4)   Are the advantages of your invention which are spelled out in the specification (i.e. function, operation, cost savings, improved efficiency, etc.) realistic and based on actual construction, tests, or field use? If no, explain.

Claims

To you and your employer, the claims are the most important part of the patent -- the only patent protection your company will get on your invention is what is covered by the claims. The language and structure of the claims may appear awkward and confusing because each claim must be written in one sentence and in a prescribed form. Accordingly, do not hesitate to call your patent lawyer and discuss the claims prior to answering the questions below. You should clearly understand the claims prior to your approval of the application.

(1)   Do the claims identify the unique features of the invention which distinguish it from earlier inventions? If no, explain.

(2)   Do the claims contain features or elements which are not actually necessary to the invention? If yes, identify them.

(3)   Are all of the important technical aspects of the invention covered in at least some of the claims?

Oath or Declaration of Inventorship

DO NOT sign the oath or declaration until the application is in a final unmarked condition AND you agree with what it says. A patent application can be stricken from the Patent and Trademark Office files if it's altered after the oath or declaration is signed. You must be satisfied that, to the best of your knowledge and belief, every statement made in the oath or declaration is true. The signing of a declaration containing willful false statements is punishable by fine, imprisonment, or both; and such willful false statements may jeopardize the application and the validity of any patent issued as a result of that application.

Before you sign the oath or declaration of inventorship and power of attorney (if you are using an attorney), answer each of the following questions:

(1)   To the best of your knowledge, are you the original, first, and sole or joint inventor of the invention claimed in the application? If "No" see your patent lawyer.

> See Part V, Chapter 1, "How And Why To Record The Important Dates," for definitions of the terms used in Items (2), (3), and (4) below.

(2)   To the best of your knowledge, was the invention known or used by others in the United States or patented or described in a "printed publication" in the U.S. or a foreign country before the date of your invention? (The date of your invention is the date you reduced the invention to practice after you conceived it.) If "Yes" see your patent lawyer.

(3)   To the best of your knowledge, was the invention patented or described in any "printed publication" in the U.S. or a foreign country more than one year prior to this application? If "Yes" see your patent lawyer.

(4)   To the best of your knowledge, was the invention in public use, offered for sale, or sold in the United States more than one year prior to this application? If "Yes" see your patent lawyer.

(5)   Have you abandoned the invention? If "yes" see your patent lawyer.

(6)   To the best of your knowledge, was the invention the subject of any foreign patent application filed by you or your legal representatives? If "Yes" see your patent lawyer.

(7)   Have you read, do you understand, and do you agree with the final, <u>unmarked</u> version of the application's specification and claims, and the declaration of inventorship? See your patent lawyer if you answer "No" to any of the conditions in this question.

Sign the documents only after you have discussed any discrepancies in the above seven questions with your patent lawyer and are satisfied that the application is complete and accurate. The signed papers and application should be returned to your patent lawyer. After the application is filed with the Patent and Trademark Office your patent lawyer should send you a copy of the application, the serial number, and the filing date of the application. If, for any reason, the application is not filed with the Patent and Trademark Office within six weeks of your signing the oath or declaration, you will be asked to sign a new oath or declaration because the Patent and Trademark Office presumes that an oath or declaration which is more than six weeks old is stale.

V.5.1 <u>Summary</u>

Even though all persons substantively involved in preparing the application have an "obligation of candor and good faith disclosure" to the Patent and Trademark Office, and the application is written by a patent lawyer or agent, patent law holds the <u>inventor</u> responsible for its contents. Making willful false statements is punishable by fine and/or imprisonment and may jeopardize both the application and the validity of any patent obtained as a result of that application. The checklist in this chapter will allow the inventor to examine the final patent application in a systematic fashion to ensure that the requirements of the law are complied with.

PART V. CHAPTER 6: WHAT HAPPENS AFTER THE APPLICATION IS FILED

V.6.1 If The Application Is Complete

When the Patent and Trademark Office first receives a patent application they will check it to see if it's complete, i.e. if it includes the specification and any required drawings. If it's not complete they won't accept it. If it is complete, they give it a sequential serial number and a filing date. (See Section V.4.9 in Part V, Chapter 4.) Then, the Patent and Trademark Office puts the application on a stack to await examination and notifies the inventor, or his representative, of the application's filing date and serial number. This entire process takes approximately one month.

The Patent and Trademark Office will not return the original papers of a complete patent application. If an applicant has not kept, but wants, copies of the application papers, the Patent and Trademark Office will supply such copies to the applicant at a cost of $9.00 per copy.

V.6.2 If The Application Is Complete But Unsatisfactory

If the application has the specification and any required drawings but is unsatisfactory for other reasons (e.g. does not include the declaration of inventorship or filing fee), the Patent and Trademark Office will still give it a filing date and a number. However, they will notify the applicant that the application is unsatisfactory and grant the applicant a specific period of time (but always six months or less) to correct the problem. The applicant can -- on his own and for a fee -- extend this time to up to six months from the date of the notice. If the applicant does not respond within six months from the date of the notice, the Patent and Trademark Office will consider the application "abandoned" and will never look at that application again. The fee charged the applicant for extending the response time escalates from $50/month to $200/month depending on the number of months the applicant takes beyond the time granted by the Patent and Trademark Office.

V.6.3 The Examination Procedure

Complete and satisfactory applications are examined in the Patent and Trademark Office by the group in charge of the technical class of inventions to which the application relates.

The examiner will thoroughly read the specification, review the claims in the application, and evaluate the invention in the manner generally described in Part V, Chapter 3, "How to Evaluate Obviousness: A Major Factor Of Patentability." During this process, the patent examiner performs a prior art search through issued United States patents, foreign patents, and technical literature. The entire examination procedure used by the Patent and Trademark Office is explained in detail in the *Manual of Patent Office Examination Procedures*, which is available in some major public libraries or can be purchased from the:

Superintendent of Documents
U.S. Government Printing Office
Washington, DC 20402

V.6.4 The Flow Of A Patent Application

Figure 7, "Route To A Patent," depicts the flow of a patent application from the initial filing to the end of the line (i.e. the granting of a patent or a lot of wasted time, money, and effort). The left side of Figure 7 shows that after filing the patent application you have two chances of getting the patent the "easy" way; i.e. the Patent and Trademark Office will grant the patent on their first or second Official Action. Failing this, and if your company believes they're legally entitled to a patent and it's worth the expense, you can appeal to higher authorities as you move down the right side of Figure 7. Keep in mind that the appeal process is expensive and can take as long as five years, especially if you go all the way to the Supreme Court! The details of Figure 7 are discussed in this chapter.

Although an inventor is permitted to act as his own attorney in patent proceedings, we strongly recommend that you have a skilled, registered (see Part II) patent lawyer or agent assist in the prosecution of the application -- the rules for correctly prosecuting patent applications are numerous, detailed, and complex.

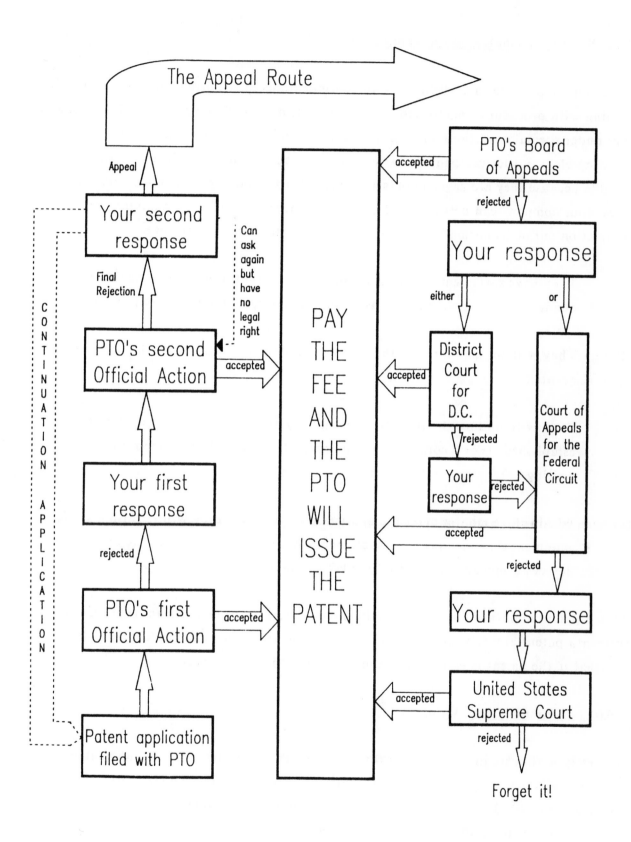

Figure 7: Route To A Patent

V.6.5 When Can You Hurry An Application?

The Patent and Trademark Office examines applications in the order they are filed or in keeping with procedures established by the Commissioner of Patents and Trademarks; and only under special conditions will a patent application be examined before its normal turn. For example, say your company wants to have their application examined as quickly as possible because they are eager to obtain a patent before starting to manufacture. In these cases, your company can petition the Patent and Trademark Office to examine the application out of its normal order, but may have to state under oath that:

(1)   They have sufficient capital and facilities to manufacture the invention in quantity. (If you were acting as an individual you would have to prove this.)

(2)   They will not manufacture the invention unless it's certain that the patent will be granted.

(3)   They will produce the invention in quantity as soon as patent protection has been established. (For a corporation, the board of directors has to commit to this in writing.)

(4)   Within three months after the application is allowed, they will furnish a sworn statement showing: (a) how much money has been expended, (b) the number of devices manufactured, and (c) the labor employed.

Also, a patent application may be examined out of normal order if it can be proven that there is a potential infringement threat to your company which can cause irreparable harm and that if the patent is issued your company can stop the infringer.

V.6.6 The First Official Action

Generally within six to eighteen months after a patent application is filed, the patent examiner will send the applicant or the applicant's representative the first written notification of his decision, i.e. his first "Official Action." If you're lucky he will write that the application is allowed, i.e. the claims are patentable. Then, upon payment of the appropriate fees, the patent is issued.

Unfortunately, more often than not the first Official Action will say that some or all of the claims have been rejected (i.e. the claims do not meet the legal requirements for patentability). The examiner will reject all of the claims if he considers the invention to be totally unpatentable. If he considers only certain claims to be unpatentable, he will reject only those claims and indicate that the other claims are, or could be with amendment, patentable. The first Official Action will give reasons for the rejections, and may include information about prior art which may help you decide if you should continue with the prosecution of the application. (See Section III.5, "Prior Art: A Major Factor Of Patentability," in Part III.)

Patent law says that only one patent can be issued for one invention. Consequently, if the Patent and Trademark Office decides that you have claimed two or more inventions in one patent application, they will not take further action until you make a choice limiting the application to only one of the inventions. That's the bad news. The good news is that you may file for the non-elected invention (i.e. the "leftover" invention) by filing a second application; and if you file the second application while the first application is pending, you will be entitled to the benefit of the filing date given to the first application. The "leftover" invention usually ends up with the same specification and drawings but with different claims than the original application.

The Language And Rules Of An Official Action

In an Official Action, the examiner uses such terms as "lack of novelty," "anticipation," and "fully met" when rejecting claims which he feels are completely known from the prior art. Further, the examiner uses such terms as "unpatentable over" and "obvious in view of" when rejecting claims which are novel but not sufficiently different to justify their acceptance. (See Part V, Chapter 3, "How To Evaluate Obviousness: A Major Factor Of Patentability.")

The commissioner can set any time limit, within patent law guidelines, to respond to an Official Action. Usually, you're given three months (30 days is allowed for relatively minor items) but always less than six months to respond. You can -- on your own and for a fee -- extend this time to up to six months from the date of the Official Action. The fee charged for extending the response time escalates from $50/month to $200/month depending on the number of months you take beyond the time granted by the Patent and Trademark Office. Personal interviews with examiners are possible; but interviews or not, you still need to respond in writing within the required time. Decisions made by the Patent and Trademark Office are based solely on what is written, not what is said.

If a response to an Official Action is not filed within six months, the Patent and Trademark Office will consider the application to be abandoned. The Patent and Trademark Office may reinstate an abandoned application when the failure to respond to an Official Action was unavoidable, such as the illness of an inventor delaying the response. To get an abandoned patent application reinstated, you must usually pay a fee and petition the Commissioner to reinstate the application.

V.6.7 Your Response To The First Official Action

If the first Official Action is negative in any respect, and you still want a patent, a written request must be sent to the Patent and Trademark Office asking that the application be reconsidered. This request must be a sincere attempt to bring the case to final action. The request can't simply say that the examiner has erred, but must specifically point out the supposed errors in the examiner's decision, addressing every objection and rejection stated in the first Official Action. You can amend or change the claims but you must clearly point out why you believe the claims are patentable. New claims can be added to the application but you may have to pay additional fees.

V.6.8 The Second Official Action

After you respond to the first Official Action, the Patent and Trademark Office will examine the application a second time and issue a second Official Action. If the claims are now found to be patentable, you only need to pay the fee and the patent will be issued.

If the application is objected to for reasons which don't relate to the patentability of the claims (e.g. objections or corrections to the drawing or specification) the examiner will probably not make the second Official Action the "final" rejection, allowing you to petition the Commissioner of Patents and Trademarks to review the application. (This possibility is not shown in Figure 7.)

However, if the examiner rejects the claims, the examiner will probably make the second Official Action the final rejection, leaving you with three options:

#1   You can ask the examiner to review the application again, and/or you can either

#2   File a "continuation" application, or

#3   Appeal the rejection to the Patent and Trademark Office.

You can't carry out both Options #2 and #3.

The rules for responding to an Official Action were described earlier in this chapter in "The Language And Rules Of An Official Action." Applying those rules here (i.e. responding to a final rejection) means Option #1 does not buy you time, and unless you want to abandon the application you must notify the examiner in writing that you have selected either Option #2 or #3.

V.6.9 Option #1: Requesting Another Examination

If the claims have been rejected twice (i.e. final rejection) you can request that the examiner review the application again, but the examiner is not legally obligated to do so. Even if he does agree to review the application, changes to the claims are not usually permitted. This means that to have any possibility of success, you must either eliminate each rejected claim or present a strong argument as to why the claim should be allowed. If at the end of your presentation the examiner feels that enough evidence exists so that the invention may be patentable, he will re-examine the application.

It must be emphasized that the examiner does not have a legal responsibility to review the application if the second Official Action is the final rejection. In fact, even if the examiner does agree to consider your arguments, the period in which you have to select either Option #2 or #3 is not extended.

V.6.10 Option #2: Filing A Continuation Application

You can file a "continuation application" if the second Official Action was the final rejection and you elect not to appeal the rejection per Option #3. A continuation application is really a new application which requires another filing fee, but if it's filed before the second Official Action response period expires, and the original application is specifically referred to, the subject matter which is common to the original application will carry the earlier filing date and any new subject matter will get the later filing date. In these cases the application has more than one date, which becomes important when determining prior art.

V.6.11 Option #3: The Appeal Route

If: (a)   The second Official Action is the final rejection, and
       (b)   You elect not to file a continuation application of Option #2 above, and
       (c)   You find that the lower courts made an error regarding the law, or application of the law,

then you have the right to follow the appeal route shown on the right side of Figure 7. You start this activity by filing a report to support your position and paying the required fee. This gets you an oral hearing by the Board of Appeals in the Patent and Trademark Office. The Board of Appeals is made up of the Commissioner of Patents and Trademarks, the Assistant Commissioners, and no more than fifteen examiners-in-chief. However, each appeal is normally heard by only three members of the Board of Appeals.

The patent is yours if you convince the Board of Appeals, but if they reject your application you have the choice of appealing to either the Court of Appeals for the Federal Circuit or filing a civil action against the Commissioner of Patents and Trademarks in the District Court for the District of Columbia. You can't do both.

*   If you decide to appeal to the Court of Appeals for the Federal Circuit, the court will review the record made in the Patent and Trademark Office and may uphold or reverse the action taken by the Board of Appeals. If the Court of Appeals accepts the application, you will be granted a patent but if they reject the application, your only remaining alternatives are to give up or take your case to the United States Supreme Court.

*   If, on the other hand, you elect to file a civil action against the Commissioner of Patents and Trademarks in the District Court, that court will make a decision on whether you are entitled to the patent. If they reject your plea, you can take your case to the Court of Appeals for the Federal Circuit; and if the Court of Appeals turns you down, your last chance is the United States Supreme Court.

As with most other legal issues, the United States Supreme Court does not have to hear your case, and most likely it's all over if they do hear your case and turn you down.

Usually you are entitled to only one hearing at each level of appeal but there are exceptions. Although such petitions are rarely granted, you can petition the Court of Appeals for the Federal Circuit or the Supreme Court for another hearing.

V.6.12 After The Patent Is Granted

If and when the claims are found to be patentable, the good news will be sent to the applicant or his representative. Then, an issue fee of $560 for utility patents and $200 for design patents is due to the Patent and Trademark Office within three months from the date of the notice. (See Section III.2, "What Can Be Patented?" in Part III for the definitions of utility and design patents.) As with the application filing fees and maintenance fees, the issue fees required from individual inventors, small companies, and non-profit organizations are one-half those required from large companies.

After the issue fee is paid, the patent is granted by the Patent and Trademark Office. On the day the patent is granted a summary of the patent is published in the *Official Gazette of the U.S. Patent and Trademark Office* (see Part V, Chapter 10) -- making the specification and drawings available to the public. The original copy of the patent is mailed to the inventor or his representative. The patent is issued in the name of the United States of America under the seal of the Patent and Trademark Office. The patent is signed by the Commissioner of Patents and Trademarks; or it has the Commissioner's name on it attested to by an official of the Patent and Trademark Office.

V.6.13 How To Correct An Issued Patent

Even though a patent is issued, it may still be "inoperative" and/or "invalid." A patent may be inoperative if the claims fall short of covering everything the owner is entitled to, or if its specification or drawing doesn't completely describe the invention. A patent is invalid if the invention is the same as, or obvious from, prior art; that is, a patent is invalid if the Patent and Trademark Office was not aware of, or did not properly evaluate, the prior art. In any event, once the patent is issued it's the patent owner's responsibility to make certain that the patent is both operative and valid; and potential competitors are wise to "let sleeping dogs lie" until they are ready to challenge the patent.

If the patent is completely or partly inoperative or invalid, and these errors were made without any deceptive intention, then a process called "reissue" can be used to correct the problem. The patent owner must file a new application for the same invention as described in the original patent, surrender the original patent to the Commissioner of Patents and Trademarks, and pay a full filing fee and full fees for each new independent and dependent claim. The reissue application will be examined (in the same manner as described above for patent applications) and a reissue patent will be granted for the remainder of the term of the original patent. The original patent will be returned if the reissue is refused.

Applications for reissue of a patent can be filed at any time during the term of the patent, but must be filed promptly after the problem is discovered. However, if the problem to be corrected is that the claims cover less than the invention was entitled to, then the owner would be seeking a "broadening" of the patent protection, and the application for reissue must be filed within two years of the issue date of the original patent.

V.6.14 How To Challenge A Patent Without Going To Court

The Patent and Trademark Office may issue patents to others for inventions which you believe are unpatentable and it's to your company's benefit to have the patents eliminated. Elimination of issued patents is possible via a procedure called "re-examination."

Anyone can request re-examination of a patent by:

(a)   paying a $1,770 fee, and

(b)   requesting that the Patent and Trademark Office re-examine the patentability of the invention, and

(c)   either giving the Patent and Trademark Office additional pertinent prior art which you believe makes the invention unpatentable, or arguing that the Patent and Trademark Office made a mistake regarding the referenced prior art which was considered during the original examination.

Once you carry out Steps (a), (b), and (c) you will not be allowed to participate further in the re-examination proceedings.

The Patent and Trademark Office is not required to re-examine a patent unless the request for re-examination raises new questions of patentability; but if they refuse to re-examine the patent they will refund 80% of the fee. If they do re-examine the patent they will re-examine the patent's claims in light of the new prior art you cited or your arguments that a mistake was made in granting the patent in the first place. The patent owner can (and most probably will) participate in the re-examination and has the right to amend the claims, add new claims, or refuse to do either. The Patent and Trademark Office can (1) find the original claims patentable, or (2) issue a patent with amended claims, or (3) reject all of the original and amended claims in the patent.

The potential for re-examination exists during the entire term of a patent. A patent can be requested to be re-examined by any number of different persons, or by any number of times by the same person, and as many times as is justified by new discoveries of prior art. This means that your company's patents are also subject to re-examination, and if re-examination is granted on one of your company's patents, your company will be involved in the re-examination just as if the patent were a newly filed application.

V.6.15 Summary

The Patent and Trademark Office won't accept a patent application unless it includes the specification and any required drawings. If the application does include those parts but is deficient for other reasons, the Patent and Trademark Office will give the applicant a specific period of time to correct the problem; and, after the problem is corrected, they will examine the application for patentability when its turn comes up, which is usually six to eighteen months after the filing date. Then they will issue their first "Official Action" notifying the applicant of the invention's patentability. If the invention is patentable, the issue fee must be paid and the patent will be granted. The patent then becomes open to the public and a copy is sent to the applicant.

Unfortunately, not all inventions are found to be patentable the first time around. If the first Official Action is negative and you want the application reconsidered, you must -- within a given time period -- make the necessary changes to the application and re-submit the application along with a written request for re-examination and a second Official Action. If the second Official Action is another rejection, then again you only have a specific period of time in which to act. If the rejection is not the "final" rejection, you can petition the Commissioner of Patents and Trademarks to have the application re-examined. If the rejection is the "final" rejection, you can (a) try to informally convince the examiner of your position; and/or choose to either (b) submit a "continuation" application, or (c)

appeal the rejection to the Patent and Trademark Office. You can't do both (b) and (c) --
and you can appeal only if the Patent and Trademark Office has made an error of law or
the application of the law. If you choose to file a "continuation" application, and the
patent is granted, all or part of the continuation application retains the benefit of the first
application's earlier filing date. If you select the appeal route and you lose the first round
(i.e. an appeal to the Patent and Trademark Office) you can take your case to the courts. If
the courts find your invention to be patentable, you will be granted a patent. If, after the
patent is issued, you discover that the patent is deficient, it's possible for you to correct
these deficiencies. In fact, any patent can be challenged by anyone at anytime during the
patent's life.

PART V. CHAPTER 7: AVOIDING PATENT INFRINGEMENT

Inherent in the patent system is the very real possibility of patent infringement -- Either your company infringing on another's patent or another infringing on your company's patent. You should avoid infringing on the patents of others not only because the law says that it's your duty to avoid infringement but also because infringing on another's patent can be very expensive.

V.7.1 What's Patent Infringement?

You infringe on another's patent when you -- without the patent owner's permission -- make, use, or sell an item or use a process which includes the owner's patented invention while his patent is still in force. What is protected by his patent is determined by the language of his patent's claims. Usually, if what you're making, using, or selling is not covered (i.e. is not exactly described or is not equivalent from a structural and functional standpoint) by **ANY** of the patent's claims you are not infringing. Obviously, the same argument applies to the possibility of others infringing on your patents.

In order to better explain the concept of patent infringement, let's refer to Figure 1 and its related discussions in Sections III.5, "Prior Art: A Major Factor Of Patentability," and III.6, "Patent Rights," in Part III. In particular, let's examine the bicycle wheel patent owner's situation. The bicycle wheel patent owner would be infringing on the Flintstone wheel patent if he made, used, or sold his bicycle wheel without first getting permission from the owner of the Flintstone wheel patent. Why? Because a bicycle wheel can not be built without practicing some of the same technology already covered by the Flintstone wheel patent.

All of this can be seen in Figure 1 itself. As depicted by the "bicycle wheel" dotted block outside of the pyramid, the bicycle wheel patent was granted because its patent application claimed technology which was, at the time, not included in any prior art reference including the Flintstone wheel. However, to practice the bicycle wheel patent requires also practicing the invention patented by the Flintstone wheel patent. (This is shown in Figure 1 by the "bicycle wheel" dotted block extending into the pyramid.) Therefore, the Flintstone wheel patent is infringed if the bicycle wheel is built without prior permission from the owner of the Flintstone wheel patent.

In our short example, there was only the Flintstone patent to contend with and no need to get other permission because the other prior art required to build a bicycle wheel was not protected by the patent laws when the bicycle wheel patent application was filed. In real life, however, several patents may be infringed at one time. In fact, if we expanded our example above to include all of Figure 1, we would find that the automobile wheel patent owner would be infringing if he built an automobile wheel without first getting permission from both the Flintstone wheel patent owner and the bicycle wheel patent owner.

It's important to realize that even though the Patent and Trademark Office may grant a patent for an invention, their decision does not, in any way, imply that if the invention is made, used, or sold it will not infringe patents held by others. The Patent and Trademark Office is interested only in (1) if the patent application meets the legal requirements, and (2) if the technology claimed in the application is novel, useful, and unobvious to those skilled in the art. This means that it's possible -- in fact, it frequently happens -- that a patent is granted even though it relates to an invention which, if it were made, used, or sold, would infringe on a previous, unexpired patent(s); and that the owner of the new patent cannot practice his invention without permission from the owner(s) of the earlier patent(s).

A Quick Summary Of Infringement

Patentability and infringement are two entirely different legal issues, and the Patent and Trademark Office's jurisdiction extends only to patentability. They do not have the authority to become involved in infringement questions and will not report any infringement they may discover to any organization. If a patent application claims technology which is different than prior art, the Patent and Trademark Office will grant the patent and leave the discovery and consequences of infringement up to the patent owners. The patent right only prevents others from practicing the technology claimed in the patent, but does not give anyone (including the patent owner) the right to practice the invention. Infringement comes about because the practicing of an invention often requires practicing technology protected by earlier patents.

V.7.2 Patent Marking

The primary purpose of patent marking is to provide notice to others of the potential for infringement if they copy some or all of a patented product.

Patent protection does not start until the patent is granted, and any notification that a patent is pending (e.g. "Patent Applied For" or "Patent Pending") has no legal effect but only serves to inform others that an application for a patent has been filed in the Patent and Trademark Office. Once a patent has been granted, however, the patent owner or anyone who (with the permission of the patent owner) makes, uses, or sells the patented products is not only permitted to mark the articles with "Patent" or "Patented" and the patent number, but the patent owner is provided with the maximum legal benefit. Many manufacturers do not put patent markings on their patented articles and, although they do not lose their patent, they are not taking full advantage of the patent laws.

The law says that the patent owner or anyone who manufactures a product under the patent (e.g. a licensee) may not recover damages from an infringer unless the infringer was notified of the infringement and continued to infringe after the notice. Even then, no damages can be collected for the infringement period prior to the infringer being notified.

On the other hand -- since a patent marking is legal notice of patent protection -- if the product is properly marked from the beginning the patent owner can recover damages from the time that the first infringement took place, or for a period of six years before the filing of the complaint, whichever is less.

Patent law says that you can't falsely mark an article in an attempt to scare off others from producing some or all of the same product. This means you can't legally mark an article as patented if a patent has not been granted; nor can you legally mark an article with statements which imply that a patent is pending when, in fact, a patent is not pending. Currently, the law allows for a maximum penalty of $500 for every such offense, and any person may sue for the penalty but is required to split the fine with the United States Government.

V.7.3 Who Is Responsible For Avoiding Infringement?

Patent law states clearly that:

(1)   It's the inventor's responsibility to determine whether his invention can be made, used, or sold without infringing upon the patent rights of others; and

(2)   If the potential infringer discovers that his invention would infringe the patent rights of others the potential infringer has a <u>duty</u> <u>to</u> <u>take</u> <u>positive</u> <u>steps</u> to avoid infringement. Or, as it is written by judges in patent cases, the inventor has an "affirmative duty" to avoid infringement.

Patent lawyers can help detect, assess, and avoid infringement; in fact, the courts have held that <u>obtaining</u> <u>and</u> <u>following</u> competent legal advice in infringement matters are required as part of the affirmative duty to avoid infringement.

V.7.4 Who Is Responsible For Detecting Infringement?

While it is the potential infringer's responsibility to <u>avoid</u> infringement, the responsibility for <u>detecting</u> infringement lies entirely with the owner of the patent that may be infringed upon.

Patents are not self-enforcing and there is no government agency responsible for detecting infringement or enforcing patents; in fact, there is no established method of detecting infringement. It may be helpful to hire a clipping service to discover references in trade literature but this in no way guarantees that you will detect infringement on your patents.

V.7.5 How Are Infringement Suits Litigated?

Once one does detect infringement on their patent they can sue in the U.S. federal district courts to collect damages for past infringement and to obtain a court order preventing further infringement. In an infringement suit, the defendant may:

(1)   argue that he is not infringing,

(2)   concede that the patent covers his product but challenge the patent's validity, or

(3)   argue that both (1) and (2) are true.

After a trial in a district court the loser can appeal to the appropriate federal court of appeals -- usually the Court of Appeals for the Federal Circuit. After that, the Supreme Court may take a case if they consider the legal issues to be of sufficient importance; but it's rare for the Supreme Court to take a patent case.

The one charged with infringement may himself start the suit in a federal district court using the Declaratory Judgment Act in order to obtain an earlier judgment allowing him to get on with his business.

V.7.6   The Penalties For Infringement Can Be Stiff

The threat of patent infringement should not be taken lightly. The courts issue injunctions against further infringement and order the infringer to pay the patent holder damages which compensate for past infringement.

While it's safe to say infringement damages may be very large and are never less than a reasonable royalty, they are difficult to estimate since there are several possible methods of determining such damages. One such method is to award actual damages; actual damages can be measured by the infringer's profit on infringing sales or by the patent owner's loss of profits which resulted from the infringement. Also, if the infringer knew of the patent but didn't fulfill his affirmative duty to avoid infringement he is usually found by the court to be a "willful infringer." A willful infringer will usually have to pay damages up to three times the amount of actual damages incurred by the patent owner. Normally, each party to infringement litigation must bear its own costs and expenses, i.e. court costs and lawyers' fees. However, in exceptional cases, the court may award the patent owner (assuming he wins) an additional amount equal to his lawyers' fees. Patent infringement litigation is expensive -- lawyers' fees often exceed $100,000 and sometimes exceed $500,000.

In one recent infringement case, Smith International, Inc. was ordered by a California district court to pay Hughes Tool Co. damages of $208,000,000 for infringing on Hughes' patents on certain kinds of drill bits -- and it's been estimated that the litigation cost Smith another $10,000,000 in legal fees!

V.7.7   The U.S. Government Can Infringe

The United States Government may use any patented invention without the permission of the patent owner, but if the government does infringe, the patent owner is entitled to obtain fair compensation from the government. This governmental right falls under the right of eminent domain. This is the same right that allows governments and government agencies to obtain real estate property for highway and other public purposes provided that the owner receives fair compensation.

V.7.8 How You Can Avoid Infringement

Engineers should take reasonable steps to avoid infringing upon the patents of others. In fact, as we said earlier, if you have knowledge of a patent, you have a duty to take positive action to avoid infringing upon that patent. Failure to act in good faith is considered to be willful infringement and will probably result in increased damages.

Below is a chronological outline of events in a typical product development program which will not only help minimize the risk of patent infringement, but also gives valuable, early technical information.

I. Before You Settle On A Final Design . . .

Look at the patents of others so that you can determine the patentability of the product design, realize what applicable patents have already been issued, and get a better idea of the state of the art. Existing patents are an excellent source of technical information; usually, the valuable technical information obtained more than justifies the total cost of a search. (See Part V, Chapter 10, "How And Where To Find Patent Information.") Specifically, identify:

(a)   the work done by others in the technological area relating to the product (see if and how others have tried to solve the same problem),

(b)   old technology shown in expired patents that can be freely copied,

(c)   newer technology that can be freely copied if unpatented or avoided if patented, and

(d)   the companies or individuals working with the same technology.

During this step you eventually will need your patent lawyer's opinion on the patentability and potential infringement of the product.

Keeping in mind the existing patents, the opinion of your patent lawyer, and your newly acquired ideas you can now continue with the new product design or design modifications to avoid potential infringement.

II. After The Design Is Virtually Completed . . .

A patent infringement study should be made to identify any potential infringement when the future product is made, used, or sold. Have your patent lawyer (1) identify and evaluate unexpired patents and advise you on the risk of infringement, and (2) identify options which are available to you in order to minimize or eliminate the risk. If your company does not have an in-house patent lawyer, patent infringement studies can be expensive -- ranging from $3,000 to $5,000 for each patent that raises infringement questions. However, patent infringement and litigation are far more costly!

If it appears that the product design must be changed to avoid infringement you should consider redesign (for which your patent lawyer can provide guidance) and/or obtaining a license to practice the earlier patent(s) (see Part V, Chapter 8, "Patents As Personal Property").

III. And Finally . . .

If the patent lawyer does not find infringement in the design, or you have obtained a license to practice the patented inventions of others, you will be able to design, produce, and market the new product with a very low risk of patent infringement.

V.7.9 Summary

Patent infringement is the unauthorized practicing (i.e. making, using, or selling) of an item while the item is protected by a patent. It's often true that even though the Patent and Trademark Office grants a patent, if the invention is practiced it can still infringe on the patents of others; and it's up to the potential infringer to take positive steps to determine if he can make, use, or sell his product without infringing upon the patent rights of others. If a patent owner detects infringement, he can sue in the courts to collect damages for past infringement and to obtain a court order preventing further infringement. Infringement liability can be devastating to the infringer. Follow the steps outlined in this chapter to avoid infringing on the patents of others when developing a new product.

PART V. CHAPTER 8: PATENTS AS PERSONAL PROPERTY

A patent or patent application is personal property, and ownership and other legal rights associated with the patent or patent application can be conveyed to others. That is, the rights can be licensed (i.e. others given permission to make, use, or sell the invention), transferred to others, or willed to the heirs of a deceased owner.

V.8.1 Conveying A Patent Or Patent Application To Another

Regardless of which patent or patent application rights are conveyed to others, the transaction should be in writing and all documents should show the patent or application number, date, inventor's name, and the title of the invention. Also, the documents should be notarized by a notary public or other officer authorized to administer oaths or perform notarial acts.

Although not required by law, a formal record of the conveyance should be recorded with the Patent and Trademark Office within three months of the date of the conveyance. This is no different than recording the ownership of real estate with a local government agency -- it provides public notice of ownership and some protection to both the seller and purchaser against unscrupulous dealings by the other.

V.8.2 Licensing A Patent

When the owner of a patent "licenses" his patent he remains the owner of the patent but allows another person to make, use, and/or sell the invention. There is no such thing as a standard form of license agreement; that is, a license is a contract between the parties and may include whatever provisions the parties agree to, including the payment of royalties.

It's possible to license only part of the legal rights. For example, the patent owner can license someone the right to make the invention, license a second party the right to use the invention, and license a third party the right to sell the invention. Licensing a patent or patent application can be tricky because there are many different legal issues which can arise. We recommend that you always get your patent lawyer involved in the drafting of licensing agreements.

V.8.3 Patent Ownership Transfers (And A Warning)

When the entire ownership interest in a patent or patent application is conveyed to another party, that party becomes the sole owner and owns <u>all</u> of the rights associated with the patent or patent application. That's plain and simple. However, when only a <u>portion</u> of the ownership is conveyed to a new owner, then a "joint ownership" is established -- and joint ownership of patents or patent applications can be devastating to the unsuspecting patent owner.

To demonstrate the problem, let's say you own a patent and you're interested in selling a small part of it, say only 10%. You have two options:

Option #1:   You can sell a <u>specific</u> patent right which amounts to 10% of the total patent's rights, or

Option #2:   You can sell 10% of each and of every one of the patent's legal rights.

Option #1 means that you could, for example, sell the entire rights to only one claim of the patent, or perhaps sell the right to use the invention but not manufacture or sell it. The person who bought the 10% interest would have those ownership rights which come with owning that part of the patent, whatever they happen to be. The new owner would be entitled to exercise the rights he purchased but would have no rights to the other 90% of the patent.

On the other hand -- and here comes the warning -- if you choose Option #2 (i.e. sell 10% of every legal right associated with the patent), and unless you have a contract which specifically spells out what the buyer can and cannot do, the buyer can act as if he owns 100% of the patent, even though he owns only a very small portion (10%). He can make, use, and sell the <u>entire</u> invention for his own profit, grant licenses on the <u>entire</u> invention to others, or sell his interest or any part of his interest. He can do all this without asking you, even though you still own 90% of the patent rights!

Patent law allows title (ownership) of patents or patent applications to be sold or transferred provided it's done in writing. Sales and other transfers of title to patents or patent applications are commonly called "assignments" and transfer some or all of the ownership interest in the patent or patent application to a new owner.

A patent or patent application can be mortgaged, i.e. used as collateral for a loan, in which case the lender owns the patent rights until the mortgage has been satisfied. The mortgaging of a patent or patent application is regarded as a conditional -- but absolute -- assignment until the assignment is canceled by the parties or by a court. When the mortgage is satisfied, ownership of the patent rights transfers back to the borrower.

The ownership transfer rules discussed above apply to all assignments including mortgaging. That is, if all ownership rights are assigned, the new owner becomes the sole owner of the patent and the new owner would have the same rights as the original patent owner had prior to the assignment. If the assignment transfers less than all of the ownership interest in the patent, a joint ownership relationship is set up. We recommend that you always involve a patent lawyer in any assignments of rights in a patent or patent application and the terms of all such assignments are stated clearly in writing.

V.8.4 When *Doesn't* The Company Own The Engineer's Patent?

State laws govern when the employer owns the rights to an employee's invention and these laws vary from state to state.

In the absence of state laws to the contrary, with or without a written agreement -- and without paying the employee anything beyond his normal salary -- the company can acquire the ownership of the employee's inventions and patents if:

(1)   The employee is hired to make inventions and product improvements, i.e. it's part of the employee's job to make inventions,

(2)   The invention is related to the performance of the employee's job, and

(3)   The invention is made on the company's time using the company's tools or facilities.

This means that in the absence of a written agreement between the company and the employee or state laws to the contrary:

(a)   Inventions made by the employee either before or after his employment with the company will NOT belong to the employer.

(b)   The company will NOT own the inventions made by an employee unless that employee was hired to make inventions or improve products and processes.

(c)   The company MAY get only a "shop right" if the employee does not make the invention on company time using the company's tools or facilities. A "shop right" is a personal, non-exclusive, royalty-free, non-transferable license to practice the employee's invention. This means the employee can license others, including the company's competitors, to practice the patented invention.

To avoid misunderstandings and provide greater certainty regarding the ownership of employee inventions, before an engineering person is hired there should exist a signed and written agreement between the company and the employee spelling out just exactly who owns what intellectual property rights. Usually such agreements convey to the company the ownership of inventions, patents, copyrights, and trade secrets made during the performance of the employee's job (See Parts VI and VIII for a discussion of copyrights and trade secrets). While these agreements can vary in scope of ownership, here are the two extremes:

(1)   An employee cannot be forced to surrender the ownership of his inventions or patents to his employer. This means that any employee can negotiate a written agreement that says that he -- and not the company -- retains ownership of his inventions. As far as some of their engineering personnel are concerned, most companies would find this to be an unacceptable situation and would probably not agree to such terms. At the other extreme,

(2)   Written agreements which state that the employer will own any and all inventions including those which are not related to the employee's job, not made on the company's time, and/or not made using the company's tools and facilities go beyond the employer's legal rights and will probably not be enforceable.

V.8.5 Summary

Patents and patent applications are personal property and may be, in whole or part, licensed, transferred, or willed to others. When conveying these interests, always have a patent lawyer involved and make sure you have a written contract specifically spelling out each party's rights and obligations. Be especially careful where only a part of the ownership is being conveyed. There should always be a written and signed company-engineering personnel agreement concerning intellectual property rights. Usually such agreements convey to the company the ownership of inventions, patents, copyrights, and trade secrets made during the performance of the engineer's job.

PART V. CHAPTER 9: **FOREIGN PATENTS**

U.S. patent protection is enforceable only within the United States, its territories, and its possessions. This means that your company can't stop anyone from making, using, or selling their invention in a foreign country unless that invention is protected by the patent laws of that same foreign country. Such protection can only be obtained by applying for, and getting, individual patents in each foreign country of interest.

V.9.1 Who Can Get A Foreign Patent (And When)?

U.S. citizens (whether working in the United States or in a foreign country) are allowed to file for foreign patents. However, in some cases U.S. citizens must first obtain a foreign filing license from the Commissioner of Patents and Trademarks. A foreign filing license is required if:

(1)   A U.S. patent application will not be filed, or

(2)   The foreign patent application will be filed before a U.S. patent application is filed, or

(3)   The foreign patent application will be filed within six months after filing a U.S. application.

There is one overriding exception to the above rules:

> It's possible that the filing of a foreign patent application may not be allowed by the U.S. Government if the invention is important to the security of the United States, or has military value, or relates to sensitive technologies (such as sophisticated electronic or computer circuitry). In these cases, before one can file for foreign patents, special permission must be obtained from the Commissioner of Patents and Trademarks and other government agencies such as the Department of Commerce and/or the Department of Defense.

If a foreign filing license is required but not obtained it's possible that the U.S. patent, if issued, will become invalid; and there are potential penalties even if there is no U.S. patent.

Normally, it's not a problem to obtain a foreign filing license. In fact, the current practice is that unless a secrecy order is imposed the foreign filing license is issued when the Patent and Trademark Office gives the U.S. application a filing date and acknowledges receipt of the application (see Part V, Chapter 6). This means that under normal circumstances -- unless a secrecy order is imposed on the invention -- the inventor will receive a foreign filing license (and therefore has permission to file for foreign patents) approximately one month after he sends in his U.S. patent application. It's important to remember, however, that:

(a)   foreign patent applications can not be filed unless a foreign filing license is, in fact, issued or until six months after the filing date of the U.S. application; and

(b)   applications for foreign patents can not be made if the U.S. Government imposes a secrecy order on the U.S. application until the secrecy order is removed.

The filing of foreign patent applications requires the assistance of knowledgeable foreign patent lawyers. Unless you have knowledge and experience in the selection of foreign patent lawyers it's usually best to have the U.S. patent lawyer handle the hiring of these foreign lawyers.

V.9.2 Selecting The Countries

A company should consider getting patent protection in countries where they or their subsidiaries may want to manufacture the product; or where substantial sales of the product are expected; or where they want to license others to make, use, or sell the product; or where competitors are located. Normally, the engineering department should participate with lawyers, marketeers, and business planners in making foreign patent filing decisions.

A WORD OF CAUTION

Be choosy when selecting the foreign countries. Foreign patent protection may be valuable but foreign patents are costly. In many foreign countries it will cost $6,000 or more just to prepare and file the patent application and obtain the patent; and in some countries annual patent taxes are high. These total costs, when multiplied by the number of countries selected, can result in a very high patent bill. Foreign patent decisions should be based on the best business planning information available.

V.9.3 Foreign Patent Laws Vs. U.S. Patent Laws

While foreign patent laws differ from each other, what's important to engineering personnel is that foreign patent laws differ from U.S. patent laws. Here are the two salient differences and their significance:

(1) Whereas in the United States you have one year to file for a patent after publication or sale of an invention, in most foreign countries you can't get a patent if the invention has been publicly disclosed anywhere in the world unless you have already filed a patent application in some country. This means any public, nonconfidential disclosure of an invention prior to the date of the U.S. patent application will result in the immediate loss of patent rights in many foreign countries. Accordingly, if you have interest in obtaining foreign patents, you should contact your patent lawyer prior to making any disclosure of an invention to any person who is not in a confidential relationship with your company. (See Part VII, "Secrecy Agreements.")

(2) There is no manufacturing requirement in the United States, i.e. your company won't lose any patent rights if they don't manufacture the patented item. However, in some foreign countries a patent owner may lose some or all of his rights to the patent if the invention is not manufactured in that country within,

say, three years. In these countries, a patent owner is not required to manufacture the patented invention himself, but can satisfy the manufacturing requirement by offering to grant licenses to others to manufacture the invention in that country. Also, in some countries, if the patent is not being practiced (i.e. made, used, or sold) any person can ask the appropriate court to force or compel the patent owner to issue a license to practice the patent. (See Part V, Chapter 8 for a discussion of patent licenses.) In a few of these countries the patent can be canceled if the patented invention is not actually being manufactured in that country. The country's government (or anyone else in that country) may start a legal action to cancel the patent.

V.9.4 Foreign Patent Treaties: To Engineering Personnel, A Question of Time Vs. $'s

There is no such thing as an international patent, but there are patent treaties that standardize some patent practices throughout most countries of the world.

The two most important treaties are the Paris Convention for the Protection of Industrial Property and the Patent Cooperation Treaty. Any country of the world may become a member of either treaty (the U.S. is a member of both) by agreeing to accept the associated obligations. Further, a patent application can be filed under either treaty. The important considerations of the two treaties are similar and provide for the following standard patent practices:

(1)   That each country guarantees to the citizens of the other member countries the same rights in patent matters that are enjoyed by its own citizens,

(2)   That the patent will be issued by the patent offices of the individual countries, and

(3)   That each member country will grant a "right of priority." This means that if the invention is filed in the country within the prescribed time dictated by the treaty, the country will treat the application as if it were filed in their country on the same day as it was filed in the first member country. Hence, you have a time period in which to make foreign patent filing decisions.

Now here's what important to engineering personnel. There is a difference between the two treaties in the <u>prescribed time</u> of (3) above. The Paris Convention allows the patent application to be filed within 12 months after filing in another member country; but the Patent Cooperation Treaty extends the time to 20 months.

The additional eight months provided by the Patent Cooperation Treaty has a potentially significant advantage; i.e. it allows you more time to decide on what foreign patent protection is needed without having to plunge into having many foreign patents just to be on the safe side. The disadvantage is that it costs somewhat more to file under the Patent Cooperation Treaty. While the additional costs are relatively small in each country, it can add up if your company files many patent applications in several countries each year.

All of this means that you and your patent lawyer need to make a conscious decision for each patent application as to whether the advantage offered by the Patent Cooperation Treaty of eight additional months to make foreign patent filing decisions is worth the extra cost.

V.9.5 <u>Summary</u>

U.S. patent protection extends only to the United States, its territories, and its possessions. Consequently, there's nothing your company can do to prevent their U.S. patented invention from being made, used, or sold in a foreign country unless they also have patent protection provided by that specific country. While the cost of obtaining many foreign patents can be high, such patent protection may well be worth the cost. Foreign patent laws differ from each other and from U.S. patent laws, but most countries follow various patent treaties which provide fairness to all patent applicants. Before any U.S. citizen can apply for a foreign patent, he must either first get a foreign filing license from the Commissioner of Patents and Trademarks or wait six months after filing his U.S. patent application. However, unless the U.S. Government imposes a secrecy order on the invention, the foreign filing license is issued automatically when the Patent and Trademark Office acknowledges receipt of the U.S. patent application. Under no circumstances can foreign patents be applied for during the time the U.S. Government has a secrecy order imposed on the invention.

PART V. CHAPTER 10: HOW AND WHERE TO FIND PATENT INFORMATION

There are two main reasons why an engineering department should be interested in the information contained in <u>other companies</u>' patents:

(1)   To protect their own company against infringing on other companys' patents; and

(2)   To speed up, reduce the cost of, and generally improve the results of their product development effort.

In the United States, the Patent and Trademark Office will reveal nothing about patent applications unless they have written authority from the applicants or their authorized representatives. However, anyone can see issued patents. While this means that the newest U.S. patents available relate to technology which is approximately 2 1/2 years old, these patents still contain a <u>lot</u> of good information since they make up the lion's share of the United States' <u>best</u> technology. If you want information which is somewhat more current, you can have patent searches performed in countries where patents are published in less time. For example, in Europe patent applications are generally published 18 months after filing the application; and patents in Belgium are published six to nine months after the filing date.

V.10.1 Where To Find Every U.S. Patent

For a review of U.S. patents you can't beat the library maintained by the United States Patent and Trademark Office in Arlington, Virginia. The library, which is available to citizens of any country, contains millions of U.S. and foreign patents dating back to 1836 (and that's further back than you probably want to go)! The patents are classified into thousands of technical classes and subclasses by number, inventor, and patent owner (if the patent owner is not the inventor). The library also has technical books, papers, and periodicals which may contain information about the invention.

To use the library one only needs to go to the Patent and Trademark Office in Arlington, Virginia and identify himself to the receptionist who will then issue an entrance pass. Study facilities are provided. Copies of normal-sized patents can be purchased for $1.50

$3.00 1993

119

each, but plan on waiting a half day to pick them up. The cost of copying a "jumbo patent" (i.e. one that contains many pages) would be higher.

It's not always necessary to travel to Arlington, Virginia to take advantage of the Patent and Trademark Office's library. Sometimes it makes sense to hire a professional, Washington-based individual to search the information for you. We will talk more about professional searchers later in the chapter.

You can also purchase copies of U.S. patents by mail. The Patent and Trademark Office sells patent order coupons individually or in quantities of 50 which can be redeemed for patent copies. The coupons can be ordered from:

Commissioner of Patents and Trademarks
Washington, DC 20231

Any order should be accompanied by either a certified bank check or money order made out to the Commissioner of Patents and Trademarks.

V.10.2 How To Keep Up To Date On The Latest Patents

If you're only interested in the very latest patents there are various sources available to you, most of which do not require leaving your desk.

(1) The *Official Gazette of the U.S. Patent and Trademark Office*

U.S. patents are issued on Tuesday of each week and on that same day the Patent and Trademark Office publishes an *Official Gazette of the U.S. Patent and Trademark Office* which lists each U.S. patent issued that day. While the *Official Gazette* does not show the complete patent, the information that is given (i.e. patent number, title of the invention, name and address of the inventor, assignee, a representative claim, and part of the drawing) is usually sufficient to let you decide if you should obtain the entire patent. Figure 8 is a copy of a typical patent page from the *Official Gazette*. The *Official Gazette* also contains information on design and plant patents, trademarks, changes in patent laws and procedures, and other related information.

The Official Gazette of the U.S. Patent and Trademark Office is available by subscription from the:

Superintendent of Documents
U.S. Government Printing Office
Washington, D.C. 20402

The price is $270 a year for subscribers whose mailing address is in the U.S. and $340 a year for subscribers in a foreign country.

upper and lower laser levels including means for generating X-rays having a wavelength effective to remove pref-

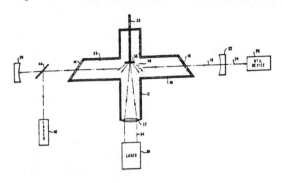

erentially inner-shell d-electrons from said atoms or ions in said target state.

4,592,065
GAS LASER EXCITED BY A TRANSVERSE ELECTRICAL DISCHARGE TRIGGERED BY PHOTOIONIZATION
Olivier de Witte, Gif sur Yvette, France, assignor to Compagnie Industrielle des Lasers Cilas Alcatel, Marcoussis, France
Filed Jun. 27, 1983, Ser. No. 507,634
Claims priority, application France, Jun. 25, 1982, 82 11172
Int. Cl.⁴ H01S 3/097

U.S. Cl. 372—83   5 Claims

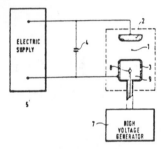

1. In a gas laser excited by a transverse electric discharge itself triggered by photoionization, said laser comprising:
  two spaced linear laser electrodes, namely a laser cathode and a laser anode, extending parallel in a longitudinal direction and facing each other, means for maintaining the space between said electrodes occupied by an active gaseous medium suitable for amplifying laser radiation when excited by said electric discharge;
  a laser capacitor having very low impedance enabling rapid discharge into said active medium to excite said medium, and having its two terminals directly connected to the two laser electrodes;
  a high energy laser charging circuit for charging said laser capacitor to an operating voltage which is less than the self-discharge voltage which would on its own cause arcs to strike across between said laser electrodes, said charge creating an operating electric field in said active medium;
  a generator of ionizing radiation for directing a pulse of trigger radiation into the active medium, after the active medium has been subjected to the operating field, said pulse being sufficiently large to trigger a uniform transverse discharge between the laser electrodes to make the medium a laser radiation amplifying medium;
  said generator of radiation comprising a high-speed, short rise time voltage generator supplying said trigger pulse in less than 10 nanoseconds, and
  wherein said high energy laser charging circuit comprises means for supplying the voltage of the laser capacitor to said laser electrodes before the ionizing radiation genera-

---

tor directs the pulse of trigger radiation into said active medium,
  whereby, a homogeneous laser discharge is obtained between the electrodes with increased energy efficiency, increased operating rate, and extended lifetime for said gas laser.

4,592,066
CONDUCTIVE BOTTOM FOR DIRECT CURRENT ELECTRIC ARC FURNACES
Eugenio Repetto; Fabrizio Marafini, both of Rome, and Francesco Tesini, Genoa, all of Italy, assignors to Italimpianti-Societa Italiana Impianti P.A., Genoa, Italy
Filed Jan. 4, 1985, Ser. No. 688,764
Claims priority, application Italy, Feb. 2, 1984, 47641 A/84
Int. Cl.⁴ H05B 7/00

U.S. Cl. 373—72   9 Claims

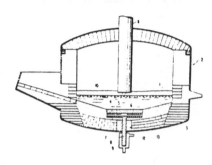

1. A conductive bottom for direct current electric arc furnaces having a hearth with a refractory bottom, the hearth comprising:
  at least one removable electrically-conductive prefabricated element in contact with the metal bath and resting on
  an intermediate part of electrically conductive granular material, resting on
  a terminal metal part, electrically insulated from the shell of the furnace, comprising a conductive metal plate with a metal rod on the face farthest from the bath, which rod passes through said refractory bottom and protrudes partially therefrom.

4,592,067
METALLURGICAL VESSEL, IN PARTICULAR AN ELECTRIC ARC FURNACE
Gerhard Fuchs, Kehl-Bodersweier; Joachim Ehle, Lautenbach, and Hans-Peter Heinzelmann, Rheinau-Diersheim, all of Fed. Rep. of Germany, assignors to Fuchs Systemtechnik GmbH, Willstatt-Legelshurst, Fed. Rep. of Germany
Filed Aug. 21, 1984, Ser. No. 642,791
Claims priority, application Fed. Rep. of Germany, Feb. 21, 1984, 3406130
Int. Cl.⁴ F27D 3/14

U.S. Cl. 373—83   13 Claims

1. A metallurgical vessel comprising,
  a main body provided with a tapping hole in the bottom thereof,
  a closure plate movable between closed and open positions outside of said tapping hole,
  a substantially horizontal actuating area connected to said closure plate for effecting lateral movement of said plate,
  a first lever pivotally supported at a first axis member mounted on the outside of said vessel main body and rotatably connected at a second axis member to said actuating arm,
  drive means connected to said first lever for pivoting the same about said first axis member and through rotation of

---

Figure 8: A Page From The Official Gazette

(2) Libraries

The libraries listed in Appendix C maintain copies of recently issued U.S. patents, and these patents are available for public inspection and copying. Normally, the libraries maintain the patents in numerical order. This makes it easy to locate a patent if you know the patent number, but very difficult to locate it if you don't know the patent number.

(3) Computerized Data Bases

Although some patent information is available from the Patent and Trademark Office via computers, the Patent and Trademark Office does not have a comprehensive computerized data base but plans to have one by 1990. However, there are private firms who are in the business of maintaining computerized data bases on patents. Their data bases generally contain the same information as does the *Official Gazette* and usually include patents issued since 1970.

For about $125 per connect hour you can access private firms' data bases by means of a personal computer and a telephone modem. Since an experienced computer operator can plug into these data bases and obtain patent information relatively quickly, the cost of a computer search is usually much less than the cost of a manual search.

Two large and commonly used data bases are Pergamon in Virginia and Dialog Information Services in California. They contain information on patents in more than 50 countries. You can obtain information about these data bases by contacting:

Pergamon ORBIT InfoLine, Inc.
8000 Westpark Drive
McLean, VA 22102
(703) 442-0900
or call toll-free (800) 421-7229

Dialog Information Services, Inc.
3460 Hillview Avenue
Palo Alto, CA 94304
(415) 858-3810

V.10.3 The Best Way To Dig Out Patent Information

How hard or easy it is to obtain patent information depends upon the kind of information you want. For example, the following information is easy to get:

*   If you want copies of a patent and you know the patent number, you may find the patent in some large public libraries (see Appendix C). If not, you can order copies from the Patent and Trademark Office for $1.50 per normal-sized patent. (Copies of jumbo patents are more expensive.)

*   If you want a list of all inventions patented by a particular inventor, you can get that list through a computer or manual search. The Patent and Trademark Office publishes annually an *Index of Inventors and Assignees*, which can be purchased from:

> Superintendent of Documents
> U.S. Government Printing Office
> Washington, D.C. 20402

This publication can also be found in large public libraries, patent law firms, and some companies. Remember also, the *Official Gazette* identifies the inventor for each patent granted that week. In some selected cases, this may be beneficial.

*   If you want a list of all patents owned by a person or company to whom patents have been assigned (i.e. they are not the inventor) you can do a computer search of assignees or do a manual search using the *Index of Inventors and Assignees* introduced above. Also, the *Official Gazette* gives this information for the patents granted that week.

*   Consult past copies of the *Official Gazette* if you want a list of all the patents issued on a particular day.

*   If you want a list of patents issued in a particular technical class or subclass, you can consult a computerized data base or the Patent and Trademark Office will provide you with that information for a fee.

* If you want to know in which technical category the Patent and Trademark Office placed a patent (see Part III), again you can use a computerized data base, or find out from the Patent and Trademark Office or the *Official Gazette*.

Getting more detailed patent information than what is itemized above becomes more difficult, more time consuming, and more expensive.

V.10.4 <u>Getting More Expensive Information</u>

* If you're interested in learning about the technical approaches which have been used in the design of a particular product you need what is commonly called a "STATE-OF-THE-ART SEARCH." A state-of-the-art search will give you information such as: (a) the work which has been done by others in the technological area, (b) old technology which can be freely copied, (c) the companies which are working in the same or similar technology, and (d) newer, patented technology that may have to be avoided. Such information is often useful to engineers in their design efforts and to lawyers in evaluating the patentability of a proposed new or modified product design.

When doing a state-of-the-art search you should, if possible, go to the Patent and Trademark Office library in Arlington, Virginia and look through the patents. It's not necessary that a patent lawyer or professional searcher help with the search, but a good professional searcher can save you time and money by directing you to the most pertinent classifications. In any event, we believe the engineering person should go to the Patent and Trademark Office and either perform the search or help in performing the search. Before going to the Patent and Trademark Office library, (a) read Section III.11, "Finding Your Way Around The Patent Document," in Part III to learn where to find what in a patent, and (b) obtain the *Manual of Patent Office Classification* and its corresponding publication, the *Index to the Manual of Patent Office Classifications*. Both of these are available from:

Superintendent of Documents
U.S. Government Printing Office
Washington, D.C. 20402

and are also available in major libraries. In addition, almost all patent lawyers keep up-to-date copies of both publications.

* To determine if an invention is patentable it is necessary to perform a "PATENTABILITY SEARCH." This is the same search that was discussed in Part V, Chapter 3, "How to Evaluate Obviousness: A Major Factor Of Patentability." A patentability search is usually best performed in the public search room at the Patent and Trademark Office in Arlington, Virginia. (See Section III.5, "Prior Art: A Major Factor of Patentability," in Part III.)

* To answer the question "Is a particular patent valid and enforceable?" a "VALIDITY SEARCH" would need to be conducted. A validity search permits a patent lawyer to determine whether a patent does, in fact, meet the legal requirements for a valid patent. (For a discussion of validity, see Section V.6.14, "How To Correct An Issued Patent," in Part V, Chapter 6.)

* If you want to know if another company's particular process, product, or design is old (i.e. the patent covering the process, product, or design has expired) and therefore it can be freely copied by anyone (including you), you need to perform a "RIGHT-TO-MAKE SEARCH." In a right-to-make search, the searcher concentrates on expired patents.

* As we discussed in Section V.7.1, "What's Patent Infringement?" in Part V, Chapter 7, the fact that an invention is patentable doesn't mean that you won't have an infringement problem. Therefore, if you have designed a product and want to identify unexpired patents which you might infringe upon if you manufacture, use, or sell your product, you would need to have an "INFRINGEMENT SEARCH" conducted. Unlike the right-to-make search discussed above where the searcher concentrates on expired patents, in an infringement search the searcher concentrates on unexpired patents.

Patent lawyers or professional searchers should definitely be used for patentability, validity, right-to-make, and infringement searches. The engineering person may want to be involved in these searches, but the patent lawyer or professional searcher should not be excluded.

126

V.10.5 Foreign Patent Searches

While the collection of patents at the U.S. Patent and Trademark Office is extensive, it does not include all foreign patents. If you want information about foreign patents you can request, or have your patent lawyer request, searches in many of the countries of the world. The most complete collections of patents exist in the European Patent Office in the Hague, Netherlands, and the national patent offices in England, Germany, and Japan.

V.10.6 A Fair Warning About Working With Searchers

Professional patent information searchers will stop searching whenever they have enough information to answer the question which was asked. Therefore, you must take great care in formulating the question you want answered.

For example, suppose the XYZ Company comes out with a new product which you want to develop for your company. If you give the searcher the obvious instructions to "find out if the XYZ Company has a patent on the product" the searcher will probably get a list of all patents owned by the XYZ Company and look for the one that may cover the new product. Then, if the searcher's answer is "Yes, they do have a patent on the product" you have a valid answer; however, if his answer is "No, they don't have a patent on the product" the answer might be accurate but can be dangerously misleading. For instance, the XYZ Company may have licensed the patent rights from the patent owner, meaning that the product is patented but not by the XYZ Company; or maybe the XYZ Company is infringing on someone else's patent. In either case, if you rely on the "no" answer and proceed to copy the XYZ Company's product, your company could spend a lot of money developing the product only to wind up with a costly infringement problem. The safe question to ask the searcher is "Does anyone have a patent on the product which is being sold by the XYZ Company?"

If you tell the searcher up front why you want certain information he can probably meet your exact needs. In fact, good searchers will ask what information you need and exactly why you need it.

V.10.7 <u>Summary</u>

The information contained in patents can help considerably in your product development activities -- not only will it help you discover and evaluate technical approaches but it will also help in avoiding patent infringement difficulties. Some patent information is easy to come by while obtaining other information is more expensive. Patent information such as names and numbers can easily be found in government publications, major libraries, and computerized data bases; and copies of issued patents can be obtained relatively cheaply and easily. However, it's usually more expensive to find out what technical design approaches others are using, if your invention is patentable, if another's patent is valid, or if your company is free to make, use, or sell their product without the risk of infringing on someone else's patent.

PART V. CHAPTER 11: **PATENT COSTS ITEMIZED**

There are various expenses involved with developing, applying for, obtaining, and maintaining U.S. and foreign patents. While individual circumstances make it difficult to estimate some costs closely, the following will provide you with a benchmark to use when recommending patent activities. The *Federal Register* published by the Patent and Trademark Office contains more information on patent fees. It can be obtained from:

Commissioner of Patents and Trademarks
Washington, DC 20231

or by calling (703) 557-1610

Patentee Or Assignee Search -- Part V, Chapter 10 $100 - $300
(paid to a professional searcher)

State-of-the-Art Search -- Part V, Chapter 10 $500 - $800
(paid to a professional searcher)

Patentability Search -- Part V, Chapter 10 $400 - $700
(paid to a professional searcher)

Patent Validity Search -- Part V, Chapter 10 $3,000 - $5,000
(paid to a professional searcher and   for each patent
lawyer for his legal opinion)   that raises
     infringement
     questions

Right-To-Make Search -- Part V, Chapter 10 $5,000 - $10,000
(paid to a professional searcher and patent lawyer)

Patent Infringement Search -- Part V, Chapter 10 $3,000 - $5,000
(*paid to a professional searcher and*   for each patent
lawyer for his legal opinion)   that raises
                                                      infringement
                                                      questions

Writing of U.S. Application -- Part V, Chapter 4 $2,500 - $5,000
(*paid to outside patent lawyer*)

Prosecuting of U.S. Application -- Part V, Chapter 6 $2,500 - $5,000
(*paid to outside patent lawyer; these costs*
can vary considerably depending upon the number
and type of rejections and appeals)

Application Filing Fee -- Part V, Chapter 4
(*paid to Patent and Trademark Office*)

    Utility patent;
       1) Individual or small company $170
       2) Large company $340

    Design patent;
       1) Individual or small company $70
       2) Large company $140

Copy Of Your Own Patent Application -- Part V, Chapter 6 $9
(*paid to Patent and Trademark Office*)

Patent Issue Fee -- Part V, Chapter 6
(*paid to Patent and Trademark Office*)

    Utility patent;
       1) Individual or small company $280
       2) Large company $560

Design patent;
    1) Individual or small company $100
    2) Large company $200

Patent Claims -- Part V, Chapter 4
(paid to Patent and Trademark Office)

Independent claims in excess of three;
    1) Individual or small company $17 each
    2) Large company $34 each

Independent and dependent claims in excess of twenty;
    1) Individual or small company $6 each
    2) Large company $12 each

All multiple dependent claims;
    1) Individual or small company $55/application
    2) Large company $110/application

U.S. Maintenance Fees -- Part III
(paid to Patent and Trademark Office;
maintenance fees are only imposed on utility
patents and not on design and plant patents)

3 1/2 years after the patent is issued;
    1) Individual or small company $225
    2) Large company $450

7 1/2 years after the patent is issued;
    1) Individual or small company $445
    2) Large company $890

11 1/2 years after the patent is issued;
    1) Individual or small company $670
    2) Large company $1,340

Surcharge for paying maintenance fees late but within the six-month grace period permitted by law -- Part III

(paid to Patent and Trademark Office)

1) Individual or small company	$55
2) Large company	$110

Fee for re-examination of issued patent -- Part V, Chapter 6 $1,770

(paid to Patent and Trademark Office)

Copies of U.S. Patents -- Part V, Chapter 10 $1.50/copy

(paid to Patent and Trademark Office)   of entire patent except in case of a "jumbo patent"

Approximate Foreign Patent Application Filing Costs -- Part V, Chapter 9

(paid to the country's government and foreign patent agents)

Canada, England	$1,500
France, Germany, Japan	$3,500
Denmark, Poland, Sweden	$4,500

Foreign Patent Taxes -- Part V, Chapter 9 $50 - $1,500/year/

(paid to the country's government)   patent/country

Summary of Total Estimated Cost of Obtaining and Maintaining a Typical U.S. Patent

Patentability Search	$400 - $700
Writing of Application	$2,500 - $5,000
Application Filing Fee	$340
Prosecution of Application	$2,500 - $5,000
Issue Fee	$560
Claims (Assume 6 independent claims)	$112
Total U.S. Maintenance Fees	$2,680
APPROXIMATE TOTAL	$9,092 - $14,392

PART VI. COPYRIGHTS

Copyright law is complex, detailed, and sometimes murky. In this section we do not try to make you an expert in copyrights but only provide the information which is useful to engineering personnel.

VI.1 <u>What Can And Can't Be Copyrighted</u>

Virtually every minute of every day you see or hear something which is protected by a copyright; for example, books, songs, movies, sports broadcasts, computer programs, engineering papers, etc., etc., etc. The items eligible for copyright protection are:

(a)   "original works of authorship" which fit into the broad categories of "literary works, dramatic works, musical works, pictorial, graphic and sculptural works, motion pictures, other audio-visual works, and sound recordings," and

(b)   "fixed in a tangible medium." This means they are expressed in a form that is either directly perceptible (such as a book) or they can be communicated with the aid of a machine (such as a computer, stereo, or VCR).

Copyright law says that, to be eligible for copyright protection, a work must satisfy both (a) and (b) above; that is, the work must be both an original work of authorship and fixed in a tangible medium. For example, an improvisational speech that has not been written or recorded is not copyrighted.

There are additional restrictions to what can be copyrighted even though the material meets the requirements of (a) and (b) above. You can't obtain copyright protection for any idea, procedure, process, system, device, method of operation, concept, principle, or discovery (all as distinguished from a description, explanation, or illustration). Further, you can't copyright such things as titles, names, short phrases, slogans, or familiar symbols or designs. Also, you can't copyright things that consist entirely of information that is common property and contain no original authorship such as standard calendars, height and weight charts, tape measures, and lists or tables taken from public documents or other common sources.

133

VI.2 <u>What Protection Copyrights Provide</u>

Copyright law says that only the copyright owner has the right to reproduce, distribute, perform, or display his copyrighted work publicly, or to base other derivative works on his copyrighted work; and only the copyright owner has the right to authorize others to do these same things. However, in spite of the exclusive rights the copyright owner has, non-owners have some limited rights to that same material. For example, non-owners can copy the copyrighted material without the owner's permission or payment to the owner as long as the use of the copied material is reasonable and not harmful to the rights of the owner. Copyright law refers to this type of use as being "fair use." Fair use of copyrighted materials is not an infringement.

"Fair use," in the eyes of the law, is when copyrighted material is used for such purposes as criticism, comment, news reporting, teaching (including multiple copies for classroom use), or research. The law considers four criteria when evaluating if a particular use of the copied material is "fair use:"

(1)   The purpose and character of the use, including if the nature of the use is nonprofit, educational or commercial;

(2)   The amount and substantiality of the material used in relation to the copyrighted work as a whole;

(3)   The nature of the copyrighted work; and

(4)   The effect of the use on the potential market or value of the copyrighted work.

Obviously the above four "fair use" criteria leave room for interpretation. However, with regard to everyday engineering works the criteria generally mean that non-owners cannot legally publish usual types of copyrighted engineering information under their name or even under the owner's name but they probably can safely make a copy of that same material for their own use.

Besides the concept of "fair use," there is another fundamental and <u>very</u> important concept engineering people need to know and understand about copyright protection. Copyright law says:

> Copyright protection extends only to the specific tangible means chosen to express ideas, thoughts, or facts but does not protect the ideas, thoughts, or facts themselves.

This means that if what you publish (even though it's copyrighted) contains a unique engineering concept, the concept itself is not protected by the copyright and <u>can be</u> legally used by others, including your competitors. If you want to protect the inventions, discoveries, or technical information contained in the publication you will have to use other forms of protection covered in this book; namely, patents, trade secrets, and secrecy agreements.

To emphasize this salient -- and often misunderstood -- concept let's look at copyright protection as it pertains to the common engineering drawing. Let's say that you distribute to persons outside of your company a design drawing of a chair (or any other functional device). If you <u>don't</u> copyright that drawing anyone can republish that drawing for any purpose they wish. If you <u>do</u> copyright the drawing no one can legally copy the drawing and use the copy beyond the limits of fair use without first getting approval from your company (assuming they own the copyright). However, under no circumstances does the copyright protect the chair itself. Whether you copyright the drawing or not, if there was not already a patent (or some other form of protection) on the design of the chair then anyone <u>can</u> legally build and sell the chair if they so wish.

VI.3 <u>What You Should Copyright (And Why)</u>

Copyright protection can be obtained on such engineering-related works such as: engineering papers and books, non-standard forms, sketches and diagrams, computer programs (including micro code), micro-circuit masks, manufacturing routings, product catalogs, service and operating manuals, brochures, advertisements, films, audio and video discs, recordings and tapes, and photographs.

Most likely, you wouldn't even think of copyrighting a simple engineering data form or sketch which is so mundane that you wouldn't care if the world used it or not. However, if you're publishing a "how to" brochure or book relating to the repair, service, or operation of one of your company's products, you should copyright that book in order to keep your competitors from republishing that same book under their name. But remember, the copyright itself does not prevent others from using the information and procedures described in the brochure or book.

In short, copyright the work if you want to protect it from being republished by others outside of your company. If you don't care if the work is republished, don't copyright. However, in no event should you give materials to persons outside of your company which contain valuable engineering concepts, inventions, discoveries, or technical information unless they are already protected by patents, trade secrets, or secrecy agreements.

VI.4 <u>When The Law Considers The Work "Published"</u>

In the next sections we emphasize that the <u>requirements</u> for copyright protection depend on whether or not the work is "published." Therefore, it's important to know that the law defines publication as:

> The distribution of copies of a work to the public by sale or other transfer of ownership, or by rental, lease, or lending. The offering to distribute copies to a group of persons for purposes of further distribution, public performance, or public display constitutes publication, but a public performance or display of a work does not of itself constitute publication. The law further defines "to the public" as distribution to persons under no explicit or implicit restrictions with respect to the disclosure of the work's contents.

All of this means that the law would probably consider your normal engineering works to be published from the time you gave them to someone who does not work for your company (unless that person was under some other specific confidentiality agreement) and would not consider the works to be published if you showed them in slide form but did not hand them out to a seminar audience.

VI.5 How To Obtain And Preserve A Copyright: An Overview

Obtaining a copyright is accomplished merely by creating an original work of authorship; and, if the work is never published, the copyright will remain in force without you doing a thing. If the work is published, however, preserving the copyright only requires two simple steps:

(a)   Each copy of the work must have the proper copyright notice; and

(b)   The notice must be in the right place on the work.

In addition, and although it does not affect copyright protection, you must deposit two copies of the work with the Copyright Office. (See Section VI.8, "Depositing The Work.")

In Section VI.6, "When Does Copyright Protection Start? How Long Does It Last?," we discuss the differences between published and unpublished works as far as copyright protection is concerned. Then, in Section VI.7, "The Format And Location Of The Copyright Notice," we detail the strict, but simple, legal guidelines which must be followed in order to preserve a copyright.

VI.6 When Does Copyright Protection Start And How Long Does It Last?

Copyright law says that:

(1)   Any work that is not published is copyrighted even though it does not carry a copyright notice. This means that the work is automatically copyrighted from the time the work is created <u>until</u> the work is published. Or, to put it another way, if you create the work but don't give it to anyone, the work is copyrighted even if you didn't put a copyright notice on it.

(2)   Any work that <u>is published</u> without a copyright notice will probably lose its copyright protection. This means that if you're giving a seminar and you hand out copyrighted materials without a proper copyright notice, then anyone can copy these materials and use them for any purpose they wish. (In some circumstances, the law provides for correcting errors or omissions of the copyright notice but it's always quicker and easier to properly copyright the materials in the first place.)

(3)   Any work that <u>is</u> <u>published</u> is copyrighted if the appropriate copyright notice is put on the work <u>before</u> it is published. This means if materials with a copyright notice are handed out at a seminar, no one can legally copy these materials without prior permission from the copyright owner unless they are used within the restrictions of "fair use."

For the purposes of this book it is sufficient to say that the copyrights for most work copyrighted since 1978 last for the life of the author plus 50 years; and when two or more authors copyright the same work, the copyright extends for the life of the last surviving author plus 50 years. In the case where the "author" is a corporation or other business entity having perpetual life (that is, it never ends or dies) copyright protection lasts for 75 years from the date the work was first published, or 100 years from the date the work was created, whichever expires first.

VI.7 <u>The Format And Location Of The Copyright Notice</u>

In most circumstances, having an enforceable copyright requires putting a proper notice of copyright on each published copy of the work. As we said earlier, the notice should be on the work at the time of first publication. One exception is mask works for microelectronic circuits. A mask work needs no copyright notice to maintain copyright protection.

The Copyright Office has regulations concerning the format and location of the copyright notice in Circular R96.201.20, *Code of Federal Regulations*, which can be obtained from:

Copyright Office

LM 455

Library of Congress

Washington, DC 20559

I. Copyright Notice Format

The purpose of a copyright notice is to inform others that the owner is claiming the work to be copyright protected, while at the same time giving the copyright owner's name and, in most cases, the first year of publication. The copyright notice differs depending on the type of work. While there are many types of works, most engineering work falls into the category of "visually perceptible" (e.g. written materials and photographs). To this end, this section deals primarily with copyright notices for "visually perceptible" works but it's important that you (a) are aware that a particular work must carry the proper notice to be qualify for copyright protection, and (b) consult your lawyer for information concerning the proper copyright notice and the location of such notice.

The copyright notice for written materials or photographs consists of three parts:

(1)   either a copyright symbol (a "c" in a circle), the word "copyright," or its abbreviation "copr."

(2)   the year of publication, and

(3)   the name of the copyright owner.

The copyright notice for microelectronic circuit masks is:

(1)   either *M* or Ⓜ , and

(2)   the name of the owner of the mask work or an abbreviation by which the name is recognized or generally known.

Mask work protection is not dependent on having a copyright notice on the work even though the work is published.

When the year of publication is required (note that it's needed for written materials or photographs but not for mask work) it should always be the year that the material was first published -- NOT when it was first conceived, recorded, or printed. If the work is published again and again without changes, the first year of publication should always appear in the copyright notice. If the work is republished with additions or changes, the notice should show the first year of publication and the year of republication. Sometimes the dates of publication and republication are

139

not clear-cut and in these situations you should consult a lawyer specializing in copyrights.

The copyright owner name to use is the name of the person or organization that owns the copyright and is entitled to the legal benefits of such ownership. Only the author or those deriving their rights through the author can rightfully claim to be the copyright owner. (See Section VI.9, "When Does The Employee Own The Copyright?") The authors of a joint work are co-owners of the copyright unless there is a written agreement to the contrary.

The following are examples of copyright notices for written materials or photographs for the first publication by a corporation or an individual:

© 1987, ABC Inc.
© 1987, John Smith
Copyright 1987, ABC Inc.
Copr. 1987, John Smith

The following is an example of a copyright notice, using the years of first and second publication, for a publication by an individual of written materials or photographs which contain both the previously published and new materials:

© 1978, 1985 John Smith

For a mask work:

M 1978, 1985 John Smith
Ⓜ 1978, 1985 John Smith.

II. Copyright Notice Location

U.S. copyright law requires that the copyright notice be shown on the publication in a manner and location which would give reasonable notice that the work is copyrighted. For example, most copyright specialists agree that when copyrighting a book the safest place to put the notice is on the title page or the page immediately following the title page. Any other location, such as the back of the book, will give a copyright infringer the opportunity to argue that the copyright notice was not in a "reasonable" place and therefore he did not see it.

Where the title page is the front cover, such as in catalogs and typical technical publications, the copyright notice should appear on either the front or back of the front cover. In loose-leaf publications or catalogs, the copyright notice should appear on each page.

As pointed out earlier, the copyright notice is permitted -- but not required -- on mask works for microelectronic circuits. If a notice is used it should be located in accordance with Circular R96.201.20, *Code of Federal Regulations*, as discussed earlier.

Although there are exceptions, the notice of copyright should appear on every copy of a published work. Because of certain aspects of copyright resulting from the Universal Copyright Convention treaty, this notice should also be used on works published or distributed outside of the United States. (See Section VI.15, "Foreign Copyright Protection.")

VI.8 Depositing The Work

Generally, the copyright owner has a legal obligation to deposit two copies of the work with the Copyright Office within three months after publishing the work in the United States. Certain categories of works are exempt from these mandatory deposit requirements, while others have reduced obligations. Where deposits are required, failure to make the deposit does not affect copyright protection but can result in fines and other penalties. Further, where deposits are required, it's possible (but not always true) that the depositing requirement can be simultaneously satisfied if and when the work is registered with the Copyright Office. (See Section VI.13, "Registering The Copyright.")

For more information on the mandatory deposit requirement, send for Circular R7d from:

> Copyright Office
> LM 455
> Library of Congress
> Washington, DC 20559

VI.9 When Does The Employee Own The Copyright?

Copyright ownership initially rests with the author or authors, but if an employee creates a copyrighted work within the scope of his employment (i.e. during the normal course of his work) the work is probably "a work made for hire" under the Copyright Act; and in these cases the company will probably own the copyright unless there is written agreement to the contrary. However, if an employee creates and copyrights a work that can be construed as being outside of his normal job activities even though the work is job-related, then the employee owns the copyright <u>unless</u> there is a written agreement (created either before or after the work is published) that the work belongs to the company.

For example, let's assume that your company designs and manufactures chairs and one of the company's engineers is asked to write a copyrighted book on chair design. In this case, your company is probably the owner of the copyright. However, if that same engineer takes it upon himself to write that same book <u>but the company did not ask him to do it</u>, the owner of the copyright is the <u>engineer</u> unless there is a written agreement, either before or after the book is published, assigning the copyright to your company.

In the case of an outside agency, e.g. consultants, if you don't specifically employ them to produce the copyrighted work, they become the copyright owner unless there is a written agreement which transfers the ownership of such copyrights to your company. (See Part X, "Outside Consultants.")

VI.10 What's Copyright Infringement?

As we said earlier in Section VI.2, "What Protection Copyrights Provide," a copyright <u>does not</u> protect the ideas contained in the copyrighted material but only protects against someone using the material itself to the point of being harmful to the rights of the owner.

Thus, infringement takes place when someone uses copyrighted material beyond the extent of "fair use." Since the possibilities for potential infringement are endless, here are some basics to keep in mind:

*   The word "copying" in a copyright infringement discussion refers to <u>imitating</u> the work in a variety of ways and is not limited to just "making a copy" on a duplicating machine. This means that infringement can take place even if the work is not mechanically "copied" as, for example, on a copy machine, VCR, or computer.

*   There is no infringement if there was no copying -- even if the two works turn out to be identical. For example, there is no copyright infringement if two persons independently write identical solutions to a problem.

*   Giving credit to the copyright owner or using just portions (even small portions) of a copyrighted work does not mean that the copyright has not been infringed.

*   Oftentimes it's arguable whether a copyright has been infringed and the decision must be made by the courts.

*   One can rarely prove infringement by direct evidence since very often there are no witnesses.

*   The courts are likely to suspect infringement when there is evidence that the alleged infringer had access to (i.e. an opportunity to copy) the copyrighted work and there exists a substantial similarity between the copyrighted work and the alleged infringing material.

The damages and penalties for copyright infringement are potentially severe. The copyright owner can, in addition to recovering monetary damages and obtaining an injunction against further infringement, confiscate or have destroyed all copies of infringing materials in the possession of the infringer, and confiscate or have destroyed the machinery on which the infringement was made. Also, the infringer may be subjected to criminal fines. (See Section VI.13 for more discussion on infringement penalties.)

Before you can sue someone for copyright infringement, the copyright must be registered with the Copyright Office in Washington, D.C. and such registration requires a recording of certain information. (See Section VI.13, "Registering The Copyright.") To be on the safe side, and even though you may never register your copyright, you should record the following evidence before you publish copyrighted works which are important to your company:

(1)   The exact date (i.e. the month, day, and year) when the work was first published.

(2)   The name and address of the printer if a printer printed the work.

(3)   At least two copies of exactly what is being published. After they are published these two copies cannot be modified in any way. These copies will be needed to comply with the copyright registration requirements.

You may never detect infringement of your copyrighted material. If you do, however, your company should contact a lawyer who specializes in copyrights to assist in notifying the infringer and requesting that the infringer pay damages, whether they be monetary or otherwise.

For a current and popular example of the complexity of copyright infringement see Appendix A, "Software Protection."

VI.11 Copying U.S. Government Works

Copyright protection is not available for any work of the United States Government. Hence, U.S. Government works can be freely copied and used by anyone. Copyright law says that if your copyrighted work "consists preponderantly of one or more works of the United States Government" you must identify, in your copyright notice, either those portions of the work which are protected by the copyright law or identify those portions that constitute United States Government material. Obviously, then, the portion of your work taken from a U.S. Government work can be freely copied by anyone without your permission and without infringing on your copyright.

VI.12 How You Can Protect Against Infringement

Since it's often difficult to prove that someone has copied copyrighted material, it's common practice to include "trap lines" in copyrighted works. Trap lines are data which are not necessary to the work and do not detract from the work but which are knowingly false or relate to something which is known only to the author. If you include trap line data in your copyrighted work and the information from the trap line shows up in someone else's work, the infringer will have a hard time convincing a court that his work was prepared independently and without copying. Of course, trap lines work both ways: you can easily be caught by trap lines placed in the copyrighted works of others.

VI.13 Registering The Copyright

Even though the copyright notice itself maintains protection on the work, the copyright can also be "registered" with the Copyright Office in Washington D.C. at any time during the life of the copyright. (There is no such thing as local or state registrations of copyrights.) All it takes is a $10 fee, a completed application for copyright registration, and two copies of the work.

> Note: There are special provisions for works published in the United States which allow a single deposit of the work to satisfy both the registration requirements and the mandatory deposit requirements. (See Section VI.8, "Depositing The Work.") A special form is not required to have this dual effect -- the mandatory deposit requirement is fulfilled by the two copies sent with the application for registration and the registration fee.

Although the law doesn't require that a copyright be registered unless you are going to sue someone for copyright infringement, there are some advantages to registering:

(1)   Registering a copyright establishes a public record of the copyright claim, which helps prove that the work was copyrighted and the date it was copyrighted.

(2)   If the copyright is registered within three months after the work is published, or registered prior to an infringement of the work, the courts can award the copyright

owner statutory damages (amounts fixed by a statute or law), attorneys' fees, actual damages, and the profits made by the infringer which are attributable to the infringement. Otherwise, only actual damages and profits can be awarded.

(3)   If the copyright is registered before or within five years after publication, registration will not only provide evidence of the copyright's validity but it will also provide evidence of the facts stated in the registration certificate.

Since the Copyright Office receives more than 500,000 applications a year you will not receive an acknowledgement that the registration application has been received. However, within 90 days you will receive a certificate of registration affirming that the work has been registered; or, if the application is rejected, a letter explaining why it was rejected. If you want to know early on that the Copyright Office received your material, send the material via registered or certified mail and request a return receipt.

The Register of Copyrights has various copyright registration application forms and associated detailed instructions for different types of copyrighted work. For example, Figure 9 is a copy of Form TX and instructions for registration of non-dramatic literary works -- the application form which your department would most likely be using for registering technical papers, books, computer programs, or service and operating manuals. (Circular R61 in Appendix A provides further information on copyrighting computer programs.) Other types of application forms include: Form SE for works to be issued in successive parts such as newsletters and magazines; Form PA for works of the performing arts; Form VA for works of the visual arts; and Form SR for sound recordings. The Copyright Office supplies these and other application forms free of charge. They can be ordered by telephoning (202) 287-9100.

A copyright registration application does not have to be prepared or filed by a lawyer but can be filed by the owner of the copyright or can be filed by the owner of any exclusive right in the copyright. For the most part, the form is self-explanatory and requires only knowledge about the copyrighted work and the copyright owner. The form asks for (and you must include):

(1)   the title of the work, together with any previous or alternative titles under which the work can be identified;

(2) the name and nationality of the author or authors, and, if one or more of the authors is deceased, the dates of their deaths;

(3) if the work is anonymous or written under a pseudonym, the nationality or domicile of the authors;

(4) the year in which creation of the work was completed;

(5) if the work has been published, the date and nation of its first publication;

(6) the name and address of the person(s) or organization claiming the copyright (i.e. the claimant);

(7) if the claimant is not the author, a brief statement of how the claimant obtained ownership of the copyright;

(8) in the case of a compilation or derivative work, an identification of any pre-existing work or works that it is based on or incorporates, and a brief general statement of the additional material covered by the copyright claim being registered;

(9) in the case of a "work made for hire," a statement to this effect (see Section VI.9, "When Does The Employee Own The Copyright?"); and

(10) other information regarded by the Register of Copyrights as bearing on the preparation or identification of the work or the existence, ownership, or duration of the copyright.

FORM TX

UNITED STATES COPYRIGHT OFFICE
LIBRARY OF CONGRESS
WASHINGTON, D.C. 20559

APPLICATION FOR COPYRIGHT REGISTRATION
for a
Nondramatic Literary Work

HOW TO APPLY FOR COPYRIGHT REGISTRATION:

- **First:** Read the information on this page to make sure Form TX is the correct application for your work.

- **Second:** Open out the form by pulling this page to the left. Read through the detailed instructions before starting to complete the form.

- **Third:** Complete spaces 1-4 of the application, then turn the entire form over and, after reading the instructions for spaces 5-11, complete the rest of your application. Use typewriter or print in dark ink. Be sure to sign the form at space 10.

- **Fourth:** Detach your completed application from these instructions and send it with the necessary deposit of the work (see below) to: Register of Copyrights, Library of Congress, Washington, D.C. 20559. Unless you have a Deposit Account in the Copyright Office, your application and deposit must be accompanied by a check or money order for $10, payable to: *Register of Copyrights.*

WHEN TO USE FORM TX: Form TX is the appropriate application to use for copyright registration covering nondramatic literary works, whether published or unpublished.

WHAT IS A "NONDRAMATIC LITERARY WORK"? The category of "nondramatic literary works" (Class TX) is very broad. Except for dramatic works and certain kinds of audiovisual works, Class TX includes all types of works written in words (or other verbal or numerical symbols). A few of the many examples of "nondramatic literary works" include fiction, nonfiction, poetry, periodicals, textbooks, reference works, directories, catalogs, advertising copy, and compilations of information.

DEPOSIT TO ACCOMPANY APPLICATION: An application for copyright registration must be accompanied by a deposit representing the entire work for which registration is to be made. The following are the general deposit requirements as set forth in the statute:

Unpublished work: Deposit one complete copy (or phonorecord).

Published work: Deposit two complete copies (or phonorecords) of the best edition.

Work first published outside the United States: Deposit one complete copy (or phonorecord) of the first foreign edition.

Contribution to a collective work: Deposit one complete copy (or phonorecord) of the best edition of the collective work.

These general deposit requirements may vary in particular situations. For further information about copyright deposit, write to the Copyright Office.

THE COPYRIGHT NOTICE: For published works, the law provides that a copyright notice in a specified form "shall be placed on all publicly distributed copies from which the work can be visually perceived." Use of the copyright notice is the responsibility of the copyright owner and does not require advance permission from the Copyright Office. The required form of the notice for copies generally consists of three elements: (1) the symbol "©", or the word "Copyright", or the abbreviation "Copr."; (2) the year of first publication; and (3) the name of the owner of copyright. For example: "© 1978 Constance Porter". The notice is to be affixed to the copies "in such manner and location as to give reasonable notice of the claim of copyright." Unlike the law in effect before 1978, the new copyright statute provides procedures for correcting errors in the copyright notice, and even for curing the omission of the notice. However, a failure to comply with the notice requirements may still result in the loss of some copyright protection and, unless corrected within five years, in the complete loss of copyright. For further information about the copyright notice and the procedures for correcting errors or omissions, write to the Copyright Office.

DURATION OF COPYRIGHT: For works that were created after the effective date of the new statute (January 1, 1978), the basic copyright term will be the life of the author and fifty years after the author's death. For works made for hire, and for certain anonymous and pseudonymous works, the duration of copyright will be 75 years from publication or 100 years from creation, whichever is shorter. These same terms of copyright will generally apply to works that had been created before 1978 but had not been published or copyrighted before that date. For further information about the duration of copyright, including the terms of copyrights already in existence before 1978, write for Circular R15a.

Figure 9: Copyright Registration Form

HOW TO FILL OUT FORM TX

Specific Instructions for Spaces 1-4

- The line-by-line instructions on this page are keyed to the spaces on the first page of Form TX, printed opposite.
- Please read through these instructions before you start filling out your application, and refer to the specific instructions for each space as you go along.

SPACE 1: TITLE

- **Title of this Work:** Every work submitted for copyright registration must be given a title that is capable of identifying that particular work. If the copies or phonorecords of the work bear a title (or an identifying phrase that could serve as a title), transcribe its wording completely and exactly on the application. Remember that indexing of the registration and future identification of the work will depend on the information you give here.

- **Periodical or Serial Issue:** Periodicals and other serials are publications issued at intervals under a general title, such as newspapers, magazines, journals, newsletters, and annuals. If the work being registered is an entire issue of a periodical or serial, give the over-all title of the periodical or serial in the space headed "Title of this Work," and add the specific information about the issue in the spaces provided. If the work being registered is a contribution to a periodical or serial issue, follow the instructions for "Publication as a Contribution."

- **Previous or Alternative Titles:** Complete this space if there are any additional titles for the work under which someone searching for the registration might be likely to look, or under which a document pertaining to the work might be recorded.

- **Publication as a Contribution:** If the work being registered has been published as a contribution to a periodical, serial, or collection, give the title of the contribution in the space headed "Title of this Work." Then, in the line headed "Publication as a Contribution," give information about the larger work in which the contribution appeared.

SPACE 2: AUTHORS

- **General Instructions:** First decide, after reading these instructions, who are the "authors" of this work for copyright purposes. Then, unless the work is a "collective work" (see below), give the requested information about every "author" who contributed any appreciable amount of copyrightable matter to this version of the work. If you need further space, use the attached Continuation Sheet and, if necessary, request additional Continuation Sheets (Form TX/CON).

- **Who is the "Author"?** Unless the work was "made for hire," the individual who actually created the work is its "author." In the case of a work made for hire, the statute provides that "the employer or other person for whom the work was prepared is considered the author."

- **What is a "Work Made for Hire"?** A "work made for hire" is defined as: (1) "a work prepared by an employee within the scope of his or her employment"; or (2) "a work specially ordered or commissioned" for certain uses specified in the statute, but only if there is a written agreement to consider it a "work made for hire."

- **Collective Work:** In the case of a collective work, such as a periodical issue, anthology, collection of essays, or encyclopedia, it is sufficient to give information about the author of the collective work as a whole.

- **Author's Identity Not Revealed:** If an author's contribution is "anonymous" or "pseudonymous," it is not necessary to give the name and dates for that author. However, the citizenship or domicile of the author **must** be given in all cases, and information about the nature of that author's contribution to the work should be included.

- **Name of Author:** The fullest form of the author's name should be given. If you have checked "Yes" to indicate that the work was "made for hire," give the full legal name of the employer (or other person for whom the work was prepared). You may also include the name of the employee (for example, "Elster Publishing Co., employer for hire of John Ferguson"). If the work is "anonymous" you may: (1) leave the line blank, or (2) state "Anonymous" in the line, or (3) reveal the author's identity. If the work is "pseudonymous" you may (1) leave the line blank, or (2) give the pseudonym and identify it as such (for example: "Huntley Haverstock, pseudonym"), or (3) reveal the author's name, making clear which is the real name and which is the pseudonym (for example, "Judith Barton, whose pseudonym is Madeleine Elster").

- **Dates of Birth and Death:** If the author is dead, the statute requires that the year of death be included in the application unless the work is anonymous or pseudonymous. The author's birth date is optional, but is useful as a form of identification. Leave this space blank if the author's contribution was a "work made for hire."

- **"Anonymous" or "Pseudonymous" Work:** An author's contribution to a work is "anonymous" if that author is not identified on the copies or phonorecords of the work. An author's contribution to a work is "pseudonymous" if that author is identified on the copies or phonorecords under a fictitious name.

- **Author's Nationality or Domicile:** Give the country of which the author is a citizen, or the country in which the author is domiciled. The statute requires that either nationality or domicile be given in all cases.

- **Nature of Authorship:** After the words "Author of" give a brief general statement of the nature of this particular author's contribution to the work. Examples: "Entire text"; "Co-author of entire text"; "Chapters 11-14"; "Editorial revisions"; "Compilation and English translation"; "Illustrations."

SPACE 3: CREATION AND PUBLICATION

- **General Instructions:** Do not confuse "creation" with "publication." Every application for copyright registration must state "the year in which creation of the work was completed." Give the date and nation of first publication only if the work has been published.

- **Creation:** Under the statute, a work is "created" when it is fixed in a copy or phonorecord for the first time. Where a work has been prepared over a period of time, the part of the work existing in fixed form on a particular date constitutes the created work on that date. The date you give here should be the year in which the author completed the particular version for which registration is now being sought, even if other versions exist or if further changes or additions are planned.

- **Publication:** The statute defines "publication" as "the distribution of copies or phonorecords of a work to the public by sale or other transfer of ownership, or by rental, lease, or lending"; a work is also "published" if there has been an "offering to distribute copies or phonorecords to a group of persons for purposes of further distribution, public performance, or public display." Give the full date (month, day, year) when, and the country where, publication first occurred. If first publication took place simultaneously in the United States and other countries, it is sufficient to state "U.S.A."

SPACE 4: CLAIMANT(S)

- **Name(s) and Address(es) of Copyright Claimant(s):** Give the name(s) and address(es) of the copyright claimant(s) in this work. The statute provides that copyright in a work belongs initially to the author of the work (including, in the case of a work made for hire, the employer or other person for whom the work was prepared). The copyright claimant is either the author of the work or a person or organization that has obtained ownership of the copyright initially belonging to the author.

- **Transfer:** The statute provides that, if the copyright claimant is not the author, the application for registration must contain "a brief statement of how the claimant obtained ownership of the copyright." If any copyright claimant named in space 4 is not an author named in space 2, give a brief, general statement summarizing the means by which that claimant obtained ownership of the copyright.

Figure 9: Copyright Registration Form

PRIVACY ACT ADVISORY STATEMENT
Required by the Privacy Act of 1974 (Public Law 93-579)

AUTHORITY FOR REQUESTING THIS INFORMATION
• Title 17, U.S.C., Secs. 409 and 410

FURNISHING THE REQUESTED INFORMATION IS:
• Voluntary

BUT IF THE INFORMATION IS NOT FURNISHED:
• It may be necessary to delay or refuse registration
• You may not be entitled to certain relief, remedies, and benefits provided in chapters 4 and 5 of title 17, U.S.C.

PRINCIPAL USES OF REQUESTED INFORMATION
• Establishment and maintenance of a public record
• Examination of the application for compliance with legal requirements

OTHER ROUTINE USES:
• Public inspection and copying
• Preparation of public indexes

• Preparation of public catalogs of copyright registrations
• Preparation of search reports upon request

NOTE:
• No other advisory statement will be given you in connection with this application
• Please retain this statement and refer to it if we communicate with you regarding this application

INSTRUCTIONS FOR FILLING OUT SPACES 5-11 OF FORM TX

SPACE 5: PREVIOUS REGISTRATION

• *General Instructions:* The questions in space 5 are intended to find out whether an earlier registration has been made for this work and, if so, whether there is any basis for a new registration. As a general rule, only one basic copyright registration can be made for the same version of a particular work.

• *Same Version:* If this version is substantially the same as the work covered by a previous registration, a second registration is not generally possible unless: (1) the work has been registered in unpublished form and a second registration is now being sought to cover the first published edition, or (2) someone other than the author is identified as copyright claimant in the earlier registration, and the author is now seeking registration in his or her own name. If either of these two exceptions apply, check the appropriate box and give the earlier registration number and date. Otherwise, do not submit Form TX; instead, write the Copyright Office for information about supplementary registration or recordation of transfers of copyright ownership.

• *Changed Version:* If the work has been changed, and you are now seeking registration to cover the additions or revisions, check the third box in space 5, give the earlier registration number and date, and complete both parts of space 6.

• *Previous Registration Number and Date:* If more than one previous registration has been made for the work, give the number and date of the latest registration.

SPACE 6: COMPILATION OR DERIVATIVE WORK

• *General Instructions:* Complete both parts of space 6 if this work is a "compilation," or "derivative work," or both, and if it incorporates one or more earlier works that have already been published or registered for copyright, or that have fallen into the public domain. A "compilation" is defined as "a work formed by the collection and assembling of preexisting materials or of data that are selected, coordinated, or arranged in such a way that the resulting work as a whole constitutes an original work of authorship." A "derivative work" is "a work based on one or more preexisting works." Examples of derivative works include translations, fictionalizations, arrangements, abridgments, condensations, or "any other form in which a work may be recast, transformed, or adapted." Derivative works also include works "consisting of editorial revisions, annotations, elaborations, or other modifications" if these changes, as a whole, represent an original work of authorship.

• *Preexisting Material:* If the work is a compilation, give a brief, general statement describing the nature of the material that has been compiled. Example: "Compilation of all published 1917 speeches of Woodrow Wilson." In the case of a derivative work, identify the preexisting work that has been recast, transformed, or adapted. Example: "Russian version of Goncharov's 'Oblomov.'"

• *Material Added to this Work:* The statute requires a "brief, general statement of the additional material covered by the copyright claim being registered." This statement should describe all of the material in this particular version of the work that: (1) represents an original work of authorship; and (2) has not fallen into the public domain; and (3) has not been previously published; and (4) has not been previously registered for copyright in unpublished form. Examples: "Foreword, selection, arrangement, editing, critical annotations"; "Revisions throughout; chapters 11-17 entirely new".

SPACE 7: MANUFACTURING PROVISIONS

• *General Instructions:* The copyright statute currently provides, as a general rule, and with a number of exceptions, that the copies of a published work "consisting preponderantly of nondramatic literary material that is in the English language" be manufactured in the United States or Canada in order to be lawfully imported and publicly distributed in the United States. At the present time, applications for copyright registration covering published works that consist mainly of nondramatic text matter in English must, in most cases, identify those who performed certain processes in manufacturing the copies, together with the places where those processes were performed. *Please note:* The information must be given even if the copies were manufactured outside the United States or Canada; registration will be made regardless of the places of manufacture identified in space 7. In general, the processes covered by this provision are: (1) typesetting and plate-making (where a typographic process preceded the actual printing); (2) the making of plates by a lithographic or photoengraving process (where this was a final or intermediate step before printing); and (3) the final printing and binding processes (in all cases). Leave space 7 blank if your work is unpublished or is not in English.

• *Import Statement:* As an exception to the manufacturing provisions, the statute prescribes that, where manufacture has taken place outside the United States or Canada, a maximum of 2000 copies of the foreign edition can be imported into the United States without affecting the copyright owner's rights. For this purpose, the Copyright Office will issue an import statement upon request and payment of a fee of $3 at the time of registration or at any later time. For further information about import statements, ask for Form IS.

SPACE 8: REPRODUCTION FOR USE OF BLIND OR PHYSICALLY-HANDICAPPED PERSONS

• *General Instructions:* One of the major programs of the Library of Congress is to provide Braille editions and special recordings of works for the exclusive use of the blind and physically handicapped. In an effort to simplify and speed up the copyright licensing procedures that are a necessary part of this program, section 710 of the copyright statute provides for the establishment of a voluntary licensing system to be tied in with copyright registration. Under this system, the owner of copyright in a nondramatic literary work has the option, at the time of registration on Form TX, to grant to the Library of Congress a license to reproduce and distribute Braille editions and "talking books" or "talking magazines" of the work being registered. The Copyright Office regulations provide that, under the license, the reproduction and distribution must be solely for the use of persons who are certified by competent authority as unable to read normal printed material as a result of physical limitations. The license is nonexclusive, and may be terminated upon 90 days notice. For further information, write for Circular R63.

• *How to Grant the License:* The license is entirely voluntary. If you wish to grant it, check one of the three boxes in space 8. Your check in one of these boxes, together with your signature in space 10, will mean that the Library of Congress can proceed to reproduce and distribute under the license without further paperwork.

SPACES 9, 10, 11: FEE, CORRESPONDENCE, CERTIFICATION, RETURN ADDRESS

• *Deposit Account and Mailing Instructions (Space 9):* If you maintain a Deposit Account in the Copyright Office, identify it in space 9. Otherwise you will need to send the registration fee of $10 with your application. The space headed "Correspondence" should contain the name and address of the person to be consulted if correspondence about this application becomes necessary.

• *Certification (Space 10):* The application is not acceptable unless it bears the handwritten signature of the author or other copyright claimant, or of the owner of exclusive right(s), or of the duly authorized agent of such author, claimant, or owner.

• *Address for Return of Certificate (Space 11):* The address box must be completed legibly, since the certificate will be returned in a window envelope.

Figure 9: Copyright Registration Form

FORM TX

REGISTRATION NUMBER
TX   TXU

EFFECTIVE DATE OF REGISTRATION
. . . .
(Month)   (Day)   (Year)

DO NOT WRITE ABOVE THIS LINE. IF YOU NEED MORE SPACE, USE CONTINUATION SHEET (FORM TX/CON)

(1) Title

TITLE OF THIS WORK: **PREVIOUS OR ALTERNATIVE TITLES:**

If a periodical or serial give: Vol. . . . . . . No. . . . . . . Issue Date . . . .

PUBLICATION AS A CONTRIBUTION: (If this work was published as a contribution to a periodical, serial, or collection, give information about the collective work in which the contribution appeared.)

Title of Collective Work: . . . . Vol. . . . . . No. . . . . . Date . . . . Pages. . . . .

(2) Author(s)

IMPORTANT: Under the law, the "author" of a "work made for hire" is generally the employer, not the employee (see instructions). If any part of this work was "made for hire" check "Yes" in the space provided, give the employer (or other person for whom the work was prepared) as "Author" of that part, and leave the space for dates blank.

1

NAME OF AUTHOR: **DATES OF BIRTH AND DEATH:**

Was this author's contribution to the work a "work made for hire"?   Yes. . . . . .   No. . . . . .   Born . . . . Died . . . . (Year)   (Year)

AUTHOR'S NATIONALITY OR DOMICILE: **WAS THIS AUTHOR'S CONTRIBUTION TO THE WORK:**

Citizen of . . . . } or { Domiciled in . . . .
(Name of Country)   (Name of Country)

Anonymous?   Yes . . . . . . No . . . . . .
Pseudonymous?   Yes . . . . . . No . . . . . .

AUTHOR OF: (Briefly describe nature of this author's contribution)

If the answer to either of these questions is "Yes," see detailed instructions attached.

2

NAME OF AUTHOR: **DATES OF BIRTH AND DEATH:**

Was this author's contribution to the work a "work made for hire"?   Yes. . . . . .   No. . . . . .   Born . . . . Died . . . . (Year)   (Year)

AUTHOR'S NATIONALITY OR DOMICILE: **WAS THIS AUTHOR'S CONTRIBUTION TO THE WORK:**

Citizen of . . . . } or { Domiciled in . . . .
(Name of Country)   (Name of Country)

Anonymous?   Yes . . . . . . No . . . . . .
Pseudonymous?   Yes . . . . . . No . . . . . .

AUTHOR OF: (Briefly describe nature of this author's contribution)

If the answer to either of these questions is "Yes," see detailed instructions attached.

3

NAME OF AUTHOR: **DATES OF BIRTH AND DEATH:**

Was this author's contribution to the work a "work made for hire"?   Yes. . . . . .   No. . . . . .   Born . . . . Died . . . . (Year)   (Year)

AUTHOR'S NATIONALITY OR DOMICILE: **WAS THIS AUTHOR'S CONTRIBUTION TO THE WORK:**

Citizen of . . . . } or { Domiciled in . . . .
(Name of Country)   (Name of Country)

Anonymous?   Yes . . . . . . No . . . . . .
Pseudonymous?   Yes . . . . . . No . . . . . .

AUTHOR OF: (Briefly describe nature of this author's contribution)

If the answer to either of these questions is "Yes," see detailed instructions attached.

(3) Creation and Publication

YEAR IN WHICH CREATION OF THIS WORK WAS COMPLETED: **DATE AND NATION OF FIRST PUBLICATION:**

Year. . . . .

(This information must be given in all cases.)

Date. . . . .
(Month)   (Day)   (Year)

Nation . . . .
(Name of Country)

(Complete this block ONLY if this work has been published.)

(4) Claimant(s)

NAME(S) AND ADDRESS(ES) OF COPYRIGHT CLAIMANT(S):

TRANSFER: (If the copyright claimant(s) named here in space 4 are different from the author(s) named in space 2, give a brief statement of how the claimant(s) obtained ownership of the copyright.)

• Complete all applicable spaces (numbers 5-11) on the reverse side of this page
• Follow detailed instructions attached   • Sign the form at line 10

DO NOT WRITE HERE
Page 1 of . . . . pages

Figure 9: Copyright Registration Form

	EXAMINED BY:	APPLICATION RECEIVED:	
	CHECKED BY:		FOR COPYRIGHT OFFICE USE ONLY
	CORRESPONDENCE: ☐ Yes	DEPOSIT RECEIVED:	
	DEPOSIT ACCOUNT FUNDS USED: ☐	REMITTANCE NUMBER AND DATE:	

DO NOT WRITE ABOVE THIS LINE. IF YOU NEED ADDITIONAL SPACE, USE CONTINUATION SHEET (FORM TX/CON)

PREVIOUS REGISTRATION: **⑤** Previous Registration

- Has registration for this work, or for an earlier version of this work, already been made in the Copyright Office?   Yes . . . .   No . . . .
- If your answer is "Yes," why is another registration being sought? (Check appropriate box)
  - ☐ This is the first published edition of a work previously registered in unpublished form.
  - ☐ This is the first application submitted by this author as copyright claimant.
  - ☐ This is a changed version of the work, as shown by line 6 of this application.
- If your answer is "Yes," give: Previous Registration Number . . . . Year of Registration . . . .

COMPILATION OR DERIVATIVE WORK: (See instructions) **⑥** Compilation or Derivative Work

PREEXISTING MATERIAL: (Identify any preexisting work or works that this work is based on or incorporates.)

{ . . . .

MATERIAL ADDED TO THIS WORK: (Give a brief, general statement of the material that has been added to this work and in which copyright is claimed.)

{ . . . .

MANUFACTURERS AND LOCATIONS: (If this is a published work consisting preponderantly of nondramatic literary material in English, the law may require that the copies be manufactured in the United States or Canada for full protection. If so, the names of the manufacturers who performed certain processes, and the places where these processes were performed *must* be given. See instructions for details.) **⑦** Manufacturing

NAMES OF MANUFACTURERS	PLACES OF MANUFACTURE
. . . .	. . . .
. . . .	. . . .
. . . .	. . . .

REPRODUCTION FOR USE OF BLIND OR PHYSICALLY-HANDICAPPED PERSONS: (See instructions) **⑧** License For Handicapped

- Signature of this form at space 10, and a check in one of the boxes here in space 8, constitutes a non-exclusive grant of permission to the Library of Congress to reproduce and distribute solely for the blind and physically handicapped and under the conditions and limitations prescribed by the regulations of the Copyright Office: (1) copies of the work identified in space 1 of this application in Braille (or similar tactile symbols); or (2) phonorecords embodying a fixation of a reading of that work; or (3) both.

  a ☐ Copies and phonorecords   b ☐ Copies Only   c ☐ Phonorecords Only

DEPOSIT ACCOUNT: (If the registration fee is to be charged to a Deposit Account established in the Copyright Office, give name and number of Account.) **⑨** Fee and Correspondence

CORRESPONDENCE: (Give name and address to which correspondence about this application should be sent.)

Name . . . .

Name: . . . .

Address: . . . . (Apt.)

Account Number: . . . .

. . . .
(City)   (State)   (ZIP)

CERTIFICATION: ✱ I, the undersigned, hereby certify that I am the: (Check one) **⑩** Certification (Application must be signed)

☐ author ☐ other copyright claimant ☐ owner of exclusive right(s) ☐ authorized agent of: . . . .
(Name of author or other copyright claimant, or owner of exclusive right(s))

of the work identified in this application and that the statements made by me in this application are correct to the best of my knowledge.

☞ Handwritten signature: (X) . . . .

Typed or printed name . . . . Date . . . .

MAIL CERTIFICATE TO **⑪** Address For Return of Certificate

. . . .
(Name)

. . . .
(Number, Street and Apartment Number)

. . . .
(City)   (State)   (ZIP code)

(Certificate will be mailed in window envelope)

✱ 17 U S C § 506(e): Any person who knowingly makes a false representation of a material fact in the application for copyright registration provided for by section 409 or in any written statement filed in connection with the application, shall be fined not more than $2,500.

✱ U.S. GOVERNMENT PRINTING OFFICE: 1980: 341-278/1

Nov. 1980 — 500.000

Figure 9: Copyright Registration Form

CONTINUATION SHEET FOR FORM TX

FORM TX/CON
UNITED STATES COPYRIGHT OFFICE

- If at all possible, try to fit the information called for into the spaces provided on Form TX.
- If you do not have space enough for all of the information you need to give on Form TX, use this continuation sheet and submit it with Form TX.
- If you submit this continuation sheet, leave it attached to Form TX. Or, if it becomes detached, clip (do not tape or staple) and fold the two together before submitting them.
- **PART A** of this sheet is intended to identify the basic application. **PART B** is a continuation of Space 2. **PART C** is for the continuation of Spaces 1, 4, 6, or 7. The other spaces on Form TX call for specific items of information, and should not need continuation.

REGISTRATION NUMBER
TX   TXU

EFFECTIVE DATE OF REGISTRATION

. . . .
(Month)   (Day)   (Year)

CONTINUATION SHEET RECEIVED

Page _____ of _____ pages

DO NOT WRITE ABOVE THIS LINE: FOR COPYRIGHT OFFICE USE ONLY

(A) **Identification of Application**

IDENTIFICATION OF CONTINUATION SHEET: This sheet is a continuation of the application for copyright registration on Form TX, submitted for the following work

- TITLE: (Give the title as given under the heading "Title of this Work" in Space 1 of Form TX.)

- NAME(S) AND ADDRESS(ES) OF COPYRIGHT CLAIMANT(S): (Give the name and address of at least one copyright claimant as given in Space 4 of Form TX.)

(B) **Continuation of Space 2**

☐ **NAME OF AUTHOR:**

Was this author's contribution to the work a "work made for hire"? Yes...... No......

DATES OF BIRTH AND DEATH:
Born.......... Died........
(Year)   (Year)

AUTHOR'S NATIONALITY OR DOMICILE:
Citizen of........................ } or { Domiciled in........................
(Name of Country)   (Name of Country)

WAS THIS AUTHOR'S CONTRIBUTION TO THE WORK:
Anonymous? Yes...... No......
Pseudonymous? Yes...... No......
If the answer to either of these questions is "Yes," see detailed instructions attached.

AUTHOR OF: (Briefly describe nature of this author's contribution)

☐ **NAME OF AUTHOR:**

Was this author's contribution to the work a "work made for hire"? Yes...... No......

DATES OF BIRTH AND DEATH:
Born.......... Died........
(Year)   (Year)

AUTHOR'S NATIONALITY OR DOMICILE:
Citizen of........................ } or { Domiciled in........................
(Name of Country)   (Name of Country)

WAS THIS AUTHOR'S CONTRIBUTION TO THE WORK:
Anonymous? Yes...... No......
Pseudonymous? Yes...... No......
If the answer to either of these questions is "Yes," see detailed instructions attached.

AUTHOR OF: (Briefly describe nature of this author's contribution)

☐ **NAME OF AUTHOR:**

Was this author's contribution to the work a "work made for hire"? Yes...... No......

DATES OF BIRTH AND DEATH:
Born.......... Died........
(Year)   (Year)

AUTHOR'S NATIONALITY OR DOMICILE:
Citizen of........................ } or { Domiciled in........................
(Name of Country)   (Name of Country)

WAS THIS AUTHOR'S CONTRIBUTION TO THE WORK:
Anonymous? Yes...... No......
Pseudonymous? Yes...... No......
If the answer to either of these questions is "Yes," see detailed instructions attached

AUTHOR OF: (Briefly describe nature of this author's contribution)

(C) **Continuation of Other Spaces**

CONTINUATION OF (Check which): ☐ Space 1   ☐ Space 4   ☐ Space 6   ☐ Space 7

Figure 9: Copyright Registration Form

VI.14 Profiting From Copyrighted Work

Since a copyrighted work is a type of property (specifically, a type of intellectual property) a copyrighted work can be licensed or sold. That is, an owner of a copyrighted work can authorize others to use the copyrighted materials in return for payment. Transfers of copyright ownership are normally made by contract and, although not required by law, recording the transfer of ownership in the Copyright Office will insure that the purchaser has a formal record of his legal rights to the copyright. The Copyright Office does not have forms for such transfers.

VI.15 Foreign Copyright Protection

There is no such thing as an "international copyright" that will automatically protect works throughout the entire world; rather, copyright protection in an individual country depends on the country's laws. While some countries offer little or no copyright protection for foreign works, most countries do offer such protection under certain conditions. Further, some countries are members of international copyright treaties and conventions which greatly simplifies obtaining copyright protection in these countries.

If the country of interest is a member of one of the international copyright conventions, the work may generally be protected by complying with the copyright notice requirements of the convention. One such convention (of which the United States is a member) is the Universal Copyright Convention (the UCC). Generally, an author may claim protection under the UCC if (a) the author of a work is from a member country, or (b) if the work was first published in a member country, and (c) if the work includes the copyright notice specified by the UCC.

Since the United States is a UCC member, the proper format and placement of the U.S. copyright notice will automatically protect a work in other UCC member countries. However, if you want protection in a country which is not a member of the UCC, you should first find out what is required in the way of notice and/or registration in that country to protect works of foreign origin. This should be done before the work is published anywhere, since protection often depends on the facts existing at the time of the first publication. Even if the work cannot be protected by an international convention, protection under specific provisions of the country's national laws may still be possible.

154

Circular R38a provides a list of countries which maintain copyright relations with the United States and can be obtained from:

> Copyright Office
> LM 455
> Library of Congress
> Washington, DC 20559

VI.16 <u>When And Why You Need Copyright Legal Help</u>

Your company will need the services of a lawyer specializing in copyrights if you want to enforce a copyright against an infringer. Also, you may need copyright legal help to answer questions and give opinions about the unique specifics of copyright law; and you may even need legal help to provide you with the proper copyright notice for your specific work, or to help with copyrighting that work in other countries besides the United States. However, you <u>will</u> <u>not</u> need a copyright specialist to put the U.S. copyright notice on copyrighted work, nor will you need a copyright specialist to register or deposit your work with the Copyright Office.

VI.17 <u>Summary</u>

Almost any original literary, dramatic, musical, artistic, or other intellectual work that is expressed in a fixed and tangible form can be protected by a copyright. Many engineering materials are eligible for copyright protection. A copyright does not protect any ideas contained in the material but only protects the copyright owner from someone reproducing the material beyond the limits of "fair use." The law gives examples of "fair use" as criticism, comment, news reporting, teaching, and research; and copyright infringement takes place when someone uses the copyrighted material beyond the limits of fair use. Oftentimes it's arguable whether a copyright has been infringed; thus, a lawyer specializing in copyrights should always be involved in infringement prosecution cases. The penalties for copyright infringement can be severe.

A copyright is in effect from the time the work is created, and if the work is never published the work is copyrighted whether the work carries a copyright notice or not. However, once the work is published it is only protected if it carries the proper copyright notice. Although there are benefits to doing so, the law does not require that a copyright be registered unless you are going to sue someone for copyright infringement. However, whether or not a copyright is registered, it is mandatory for the copyright owner to deposit in the Copyright Office two copies of certain (not all) published copyrighted material for use by the Library of Congress. While failing to make the mandatory deposit of the work does not negate copyright protection, there is the potential for fines and other penalties. Copyrights last for the life of the last surviving author plus 50 years and last for 75 to 100 years when the "author" is an existing company.

PART VII. SECRECY AGREEMENTS

It's common for customers, vendors, outside inventors and others to ask engineering personnel to enter into secrecy agreements. That is, (a) to keep their technical information confidential, and/or (b) to use their information for a limited purpose, or (c) not to use their information at all. On the other hand, you -- or others in your company -- may ask the same of your customers and suppliers.

Oftentimes, such secrecy agreements are signed casually without adequate consideration of the important consequences and without legal advice or assistance. This casual -- and dangerous -- attitude towards secrecy agreements hides their legal and commercial significance.

VII.1   What's A Secrecy Agreement?

A secrecy agreement is an agreement which creates a confidential relationship with someone outside your company; and is used for preserving secrecy with respect to inventions, technical information, know-how, trade secrets, and other business matters.

VII.2   Have Outsiders Sign Your Secrecy Agreement (But Don't Reveal More Than Necessary)

A signed secrecy agreement can (and is often necessary to) create and/or preserve intellectual property protection rights. For example, disclosures made without a secrecy agreement (or other obligation of confidentiality) before filing a patent application can destroy your company's patent rights (see Part V, Chapter 1, "How And Why To Record The Important Dates"); and trade secret protection may also be lost by any unrestricted disclosure of the information (see Part VIII, "Trade Secrets").

Even though signed secrecy agreements afford your company some protection, they should still be very protective of their technical information. Your company should not disclose their technology to anyone unless the disclosure is necessary for compelling commercial reasons and only divisional and higher managers should be authorized to approve the disclosure of information to outsiders. Further, if the manager does elect to disclose the technology, your lawyer should be consulted and the disclosure should only be made under a properly drafted secrecy agreement.

VII.3 Be Careful When Somebody Wants You To Keep Their Secret

Never forget that secrecy agreements are enforceable, and injunctions may be granted and large damages awarded for not adhering to them. You should never agree to commit your company to maintain in confidence (or refrain from using) the technical information of others. If you are asked to sign such an agreement (and here we're assuming you're not "top management") refuse to sign and inform top management of the request -- only top management should be authorized to sign the secrecy agreements of others. Further, before top management signs such an agreement they should carefully weigh the pros and cons of agreeing to keep the secret. They should ask the question:

> "Is there a compelling commercial or legal need to commit to this secrecy agreement?"

Unless the answer is clearly and definitely "Yes" a secrecy agreement should be avoided. And, even if the answer is "Yes" management should consult lawyers for legal guidance.

> If your company agrees to keep secret the technical information of others, your company can be limited or prevented from using the technology even though that same technology is available to the public!

In order to provide the necessary legal services in connection with secrecy agreements your company's lawyer will require the basic information asked for in Figure 10. Fill out Part 1 of Figure 10 when you want outsiders to keep your information confidential. Complete Part 2 when outsiders want you to keep their information confidential.

<u>Figure 10: Forms For Providing The Information To Your Lawyer</u>

Part 1: Information Needed When You Want Outsiders To Keep Your Secret

1.   With what company(s) or individual(s) do you want to enter into an agreement? (If companies, state the company names, not divisions or other subunits.)

2.   What subunits (e.g. divisions) of the company(s) are involved?

3.   What are the parties' addresses?

4.   What are the parties' states/countries of citizenship or incorporation?

5.   Where is the agreement likely to be performed, i.e. in which state of the United States or foreign country?

6.   What specific information is to be provided to the parties? (Indicate the particular product(s) or manufacturing process to which the technical information relates as well as the kind of information, e.g. designs, complete sets of manufacturing drawings, test specifications, chemical process details, etc.)

7.   What will the parties do with the information they receive from you? What use will they make of it? Will they use it for experimental or commercial purposes?

8.   When do you want to give the information to the parties? If over a period of time, give the period of time.

9.   For what period of time are the parties to keep the information in confidence? (The time period should be evaluated on a case-by-case basis to meet the needs of the situation.)

10.   Has your company entered into other confidentiality agreements with other parties with respect to the same or similar information? (List parties to such agreements.)

11.   Do you know if the parties entered into other confidentiality agreements relating to the same or similar information? Give details, if known.

159

Part 2: Information Needed When Outsiders Want You To Keep Their Secret

1. Who wants you to keep their technical information secret? (If it's a company, give the company's name -- not division or other subunit.)

2. What subunits (e.g. divisions) of the parties are actually involved?

3. What are the parties' addresses?

4. What are the parties' states/countries of citizenship or incorporation?

5. Where is the agreement likely to be performed, i.e. in which state of the United States or foreign country?

6. What specific information will you receive? (Indicate the particular product(s) or manufacturing process to which the technical information relates as well as the kind of information, e.g. designs, complete sets of manufacturing drawings, test specifications, chemical process details, etc.)

7. What use are you or your company going to make of the information received from the other party? Will it be used for experimental or commercial purposes?

8. When do you expect to receive the information? If over a period of time, give the period of time.

9. For what period of time is your company expected to keep the information in confidence?

10. Has your company entered into other confidentiality agreements with other parties with respect to such information? (List parties to such agreements.)

11. Do you know if the parties entered into other confidentiality agreements relating to the same or similar information? Give details, if known.

VII.4 <u>Summary</u>

You and your company should have a strong bias against entering into agreements which limit your company's right to use any technology. On the other hand, it's desirable for your company to have a policy of keeping its own technical information and technology secret.

Therefore, you and your company's business practice should be:

(1)   To resist making any commitment to keep secret the technology or technical information of others.

(2)   To avoid disclosing technology or technical information to any outsider except under an <u>enforceable</u> secrecy agreement.

(3)   That entering into any and all secrecy agreements requires the decision and signing of top management.

(4)   To consult with a qualified lawyer before entering into any and all secrecy agreements.

PART VIII: **TRADE SECRETS**

An engineering department is often the custodian of trade secrets which can be valuable assets to a company -- valuable assets which the company will lose if the engineering personnel and others in the company are not careful.

VIII.1 <u>What's A Trade Secret?</u>

A trade secret is a formula, pattern, device, or information used to gain a business advantage over competitors who do not have the secret. Engineering information related to product designs, manufacturing processes, machines, and techniques often qualify as trade secrets. Whereas patents and copyrights are based on federal laws, trade secrets are based on individual state laws; and unlike patents, which grant the patent owner a right against all infringers, the owner of a trade secret has rights only against those who (a) have agreed (either explicitly or implicitly) not to disclose the secret, or (b) have obtained the secret by improper means. Anyone else can benefit from the trade secret information in any way they wish.

Information <u>does</u> <u>not</u> have to be the basis for a patentable invention in order to be a trade secret. In fact, the information can be a trade secret even if its owner doesn't realize that it's a trade secret. Only certain kinds of information are eligible to be "trade secrets" and the handling of that information must adhere to certain rules or the information will not, in fact, be a trade secret. What ultimately is a trade secret may vary from state to state but in general:

(1)   The information can not be readily available. That is, it must not be evident to, easily obtained by, or known to a substantial segment of the industry, and

(2)   The information must be or must have been used in the business, and

(3)   The information must provide an advantage over competitors, and

(4)   The information must be "secret."

The following is an explanation of these laws.

(1)   *The Information Cannot Be Readily Available.* Trade secrets are usually technical product design information which cannot be obtained by merely observing the product, processes, or tools used in producing the product; and the easier it is to obtain the information by "reverse-engineering" or independent research the less likely the information is to qualify as a trade secret. Matters of public or general knowledge in an industry are not considered to be trade secrets which means that the information must not be readily obtainable from literature. Further, if the information is discernible from an inspection of the product, the sale of the product destroys the trade secret status of the information.

(2)   *The Information Must Be Or Must Have Been Used In The Business.* The courts are reluctant to allow every idea to be a trade secret. This makes it difficult to enforce trade secret protection for information which may not be useful -- the information doesn't have to be used constantly or even frequently, but it must be useful information. The best proof that the information is not just a brainstorm is when the information is actually used in the business (e.g. in the design or manufacture of a product).

(3)   *The Information Must Allow For An Advantage Over Competitors.* Another requirement for trade secret status is that the information must provide an advantage over competitors who do not know or use the information. Here, it's important to realize the difference between "confidential information" and "trade secrets." Many business matters are confidential (such as the date a new product will hit the market) but only information which provides a <u>demonstrable</u> competitive advantage may be considered a trade secret.

(4)   *The Information Must Be Kept "Secret."* Trade secret status can be lost if the information is disclosed to just one person who does not have "the need to know." This doesn't mean that to be classified as a trade secret the information has to be kept "absolutely secret" but only means there must be a substantial segment of the pertinent industry who does not know the information. Information is considered "secret" if it is revealed only to employees and others (such as suppliers) as required for carrying out business -- but, as we will discuss later, you must be careful when dealing with these personnel.

Even if more than one company knows the information (e.g. they separately develop similar or identical information) each of the companies can still consider the information a trade secret -- provided that a substantial segment of the industry does not know the information.

VIII.2 What's The Life Of A Trade Secret?

Trade secret status lasts for as long as the information remains secret. For example, the formula for Coca Cola has been a trade secret for approximately 100 years and is still the basis of Coke's franchising. (Coca-Cola® and Coke® are registered trademarks of the Coca-Cola Company, Atlanta, Georgia.)

VIII.3 Licensing A Trade Secret

Trade secrets can be licensed; i.e. the trade secret owner can give another permission to use his trade secret. Generally, the licensee pays a fee or royalty for using the trade secret. Your company needs to be careful, however, when it comes to agreeing to pay for the privilege of using another's trade secret. One notable example involves the formula for a famous brand of mouthwash. Over 80 years ago, a company agreed to pay for the use of the mouthwash formula trade secret. Today, that license is still valid and that company is still paying a royalty for using the mouthwash formula even though the precise formula is available to the public and can be used by anyone else without paying royalties. (This comes under the heading of "A deal is a deal!")

VIII.4 How The Courts Determine If It's A Trade Secret

Whether or not information is a trade secret becomes important to your company when they want to:

(1)   grant others a license to use the information, or

(2)   enforce their rights with respect to the information.

In any event, your company must keep the information "secret" or they can forget about it being a trade secret. Generally, if there is a dispute as to whether or not the information is a trade secret, a court will probably rule that if your company doesn't treat the information as a valuable secret, then it's not a trade secret. The courts commonly consider the following factors in determining if information is a trade secret:

(a)   The extent to which the information is known outside of the company;

(b)   The extent to which the information is known to the company's employees and others involved with the company;

(c)   The measures taken by the company to keep the information secret;

(d)   The value of the information to the company and its competitors;

(e)   The amount of effort or money expended by the company in developing the information; and

(f)   The difficulty others would have in properly acquiring or duplicating the information.

VIII.5 How To Keep Information Secret

The following are practical measures which an engineering department and others in the company should take to protect the secrecy of valuable information. These measures can also be used as evidence that the information is a trade secret. Most of the items appear obvious, yet some may not be feasible for your company. The list is neither exhaustive nor is every item essential, but each should be given serious consideration.

(1)   Consider all information which has the potential of being a trade secret to be, in fact, a trade secret.

(2)   Formally notify the employees about the trade secret status of their work and give them periodic reminders. Disclose trade secret information to other employees (that are not directly involved) only on a need-to-know basis. That is, do not provide the information to any employee unless he must know the information in order to do his job.

(3)   Obtain a written contractual obligation of confidentiality and an agreement not to engage in the unauthorized use of trade secret information from each technical and managerial employee.

There is an implied confidentiality agreement which stems from the trust the employer is entitled to place in his employees; and, therefore, a written agreement is not absolutely essential. However, a written agreement can clearly spell out the employee's obligations so that nothing is left to interpretation by the employee or the courts. A written agreement can also put valid restrictions on an employee's post-employment use of trade secret material.

(4)   When an employee who is working with trade secret information leaves the company, the employee should be reminded of his obligation not to use the trade secrets himself or give them to others. The employee should surrender all pertinent documents and his personal files should be inspected. Also, some effort should be made to keep track of where ex-employees are working.

(5)   Prohibit or restrict access by visitors and nonessential employees to areas where trade secrets are being used. This may include the use of guards, warning or restricted entry signs, etc.

(6)   Lock up documents containing trade secret information (such as blueprints, material specifications, etc.) and control their distribution. When such documents must be issued to others they should be loaned, not given. The documents should be stamped to indicate that (a) the information is confidential, (b) copies should not be made, and (c) the document must be returned upon demand or upon completion of the job for which it was loaned.

(7)   If feasible, the trade secret process should be divided into information segments, and the segments separated among various employees or departments.

(8)   Have outside parties to whom it is necessary to disclose the trade secret information (such as suppliers, customers, and research consultants) sign a secrecy agreement obligating them to keep the information confidential and not use the information without your company's authorization. (See Part VII, "Secrecy Agreements.")

(9)   Give the trade secret information to only those who need it, <u>even</u> if other parties have agreed to sign the secrecy agreement discussed above.

(10)   Screen all company technical papers, magazine articles, advertisements, etc. prior to their publication to insure that the trade secret information is not disclosed in them.

VIII.6 <u>What Happens When "The Cat's Out Of The Bag"</u>

<u>Any</u> unrestricted disclosure of the information destroys its status as a trade secret and then the information can be freely used by anyone. In fact, trade secret status is lost even if the information is <u>wrongfully</u> taken and the taker publicly discloses the information. (Your company may be able to sue the person who publicly disclosed the information -- if they know who it is -- but trade secret status is still lost!)

In industry, however, it's more common for trade secret information to be wrongfully taken and <u>secretly</u> used by the taker or by a third party to whom the wrongful taker gives or sells it. When this happens -- and if the information is still a secret to a substantial segment of the relevant industry -- it may survive as a trade secret. However, if the third party who receives the stolen trade secret does <u>not</u> know, and has no reason to know, that the secret was stolen, the third party is under no obligation to stop using the information until he is told that he is using stolen information. If, however (and here's the bad part), before finding out that the trade secret was stolen, the third party paid for the secret or changed his business strategy because of the information he received, the third party may be able to continue to use the information without having a liability to anyone. For example, let's say that a former employee of your company gives trade secret information to a competitor without receiving your company's permission to do so. That is, the former employee wrongfully took the information. Let's also say that the competitor, unaware that he has been given stolen information, uses that information to improve his product and cut into your company's market share. There is a good chance that your company would have no legal recourse against the competitor continuing to sell his improved product.

168

VIII.7 Summary

A trade secret is secret information which provides a business advantage over competitors who do not know or use the information. Technical information about product designs, and manufacturing processes, machines, and techniques often qualify as trade secrets. If your company has information which may qualify as a trade secret, your company should treat it as a trade secret or they will never have legal recourse against <u>anyone</u> who uses it.

Some of the practical measures that should be taken to keep trade secret information secret are: treating as a trade secret all information which has potential for being a valuable trade secret; notifying employees that the information is a trade secret; disclosing trade secret information to employees only on a need-to-know basis; obtaining a written contractual obligation of confidentiality from each technical and managerial employee and from outside parties to whom it is necessary to disclose the trade secret information; controlling trade secret documents and screening other pertinent publications; and prohibiting or restricting access by visitors and nonessential employees to areas where trade secrets are being used.

PART IX: OUTSIDERS' IDEAS

IX.1 What's An Outside Disclosure?

From time to time, engineering personnel are in contact with outsiders who suggest (i.e. disclose) new products or improvements to existing products. Accepting these "outside disclosures" can result in substantial legal liability to your company as explained in an American Bar Association booklet entitled *Submitting An Idea.* The booklet can be obtained from:

> American Bar Association
> American Bar Center
> 750 North Lake Shore Drive
> Chicago, IL 60611

IX.2 How To Handle Outside Disclosures

Some companies simply refuse to review ideas from nonemployees. Their reasoning is that the risk of dispute and liability associated with these reviews is not worth the potential gain since most often the ideas submitted by outsiders are of little value to their business. Generally, these companies return the outside disclosures with a letter explaining that the materials have not been reviewed and their company policy is against accepting outside submissions.

Your company may be willing to review outsiders' ideas. If so, you should protect your employer by following certain procedures. These procedures will result in some additional paper work, but should substantially reduce the risk of expensive and time-consuming disagreements with outside inventors.

I. *Ideas Received By Mail*

    (a)   If an idea is submitted by mail do not study it; do not make copies of it; and do not correspond or discuss it with the submitter. Instead, immediately forward the materials and the envelope in which they were received to your patent lawyer (or appropriate company representative).

(b) Upon receipt of the materials, your lawyer should send the submitting inventor an appropriate letter (similar to either Figure 11 or Figure 12) together with two copies of a confidential disclosure waiver (similar to Figure 13).

(c) If the submitter signs and returns one copy of the waiver, the lawyer should return copies of the submitter's materials to you for your (and perhaps your manager's) review and preliminary evaluation.

(d) After reviewing the outside submission, the lawyer should be advised of the department's interest (or lack of interest) in the submission.

(e) The lawyer should then send a suitable letter to the submitter.

II. *Ideas Received By Telephone Or In Person*

(a) You should not -- under any circumstances -- discuss technical information or ideas with anyone who telephones or approaches your company in person. Tell the person that if he wants his idea evaluated, he must submit it in writing to your company's lawyer (or your company's representative) together with a signed copy of a confidential disclosure waiver. (Copies of the waiver, which may be similar to Figure 13, can be obtained from your lawyer.) Two copies of the waiver should be given to the submitter, and the submitter should sign and return one of the copies.

(b) If the submitter submits to your company his idea in written form along with a signed copy of the confidential disclosure waiver, the materials can be reviewed and evaluated. After evaluation, the lawyer should be advised of your department's interest (or lack of interest) in the outside submission so that the lawyer can send a suitable follow-up letter to the submitter.

<u>Figure 11: Letter To Submitter, Option 1</u>

_____

(Date)

_____

(Company and/or Division)

Proposed Submission for _____

Evaluation of _____

Thank you for submitting your idea to _____ through the above division. _____ employs engineers to improve existing products and develop new products and any idea submitted from an outside source may have already been developed or is now under development. Consequently, to avoid possible confusion or misunderstanding as to the origin of an idea, a definite understanding must be reached before _____ can agree to review ideas from persons who are not employees of the company. Accordingly, your idea has not been reviewed by any person in _____ and all materials submitted are being returned.

As a general matter, _____ prefers that you consult with a patent lawyer and obtain a patent prior to submitting your idea to _____. However, if you want to have _____ evaluate your idea prior to you obtaining a patent, _____ will review your idea but only if the attached Confidential Disclosure Waiver is accepted by you. If the Confidential Disclosure Waiver is acceptable, please sign one copy and return it to me together with the material you wish _____ to review.

Upon receipt of an unmodified signed copy of the Confidential Disclosure Waiver with no changes, your idea will be evaluated and you will be advised of _____ interest in your idea.

Sincerely,

<u>Figure 12: Letter To Submitter, Option 2</u>

_____

_____

_____

     Re:_____

This refers to _____ letter addressed to _____ concerning _____.

_____ does not accept information relating to product designs or manufacturing processes from nonemployees under circumstances which could imply an expectation or obligation of confidentiality or compensation. Accordingly, your idea has not been reviewed by any person at _____ and before _____ will review your invention, you must waive any requirement you may have for confidentiality or compensation.

Attached are copies of a Confidential Disclosure Waiver. If you sign and return one unmodified copy of the waiver, engineers from _____ will review your invention. You will then be advised as to whether _____ has an interest in the invention.

In order to review and evaluate your invention, _____ engineers will require a clear, complete, written description and sketches or drawings of the invention. If you have not previously submitted such a description or drawings, please do so when and if you return a signed copy of the waiver.

                    Sincerely,

Figure 13: Confidential Disclosure Waiver

The undersigned represents that he/she has made inventions, discoveries, and possesses information and data relating to: _____

_____

These inventions, discoveries, and corresponding information are specifically identified by the sketch(es) and written materials* which are listed herein.

The undersigned understands and agrees that he/she is making this submission to _____ without expectation of compensation, to obtain the opinion of _____ concerning the commercial merit of such inventions and the possible interest of _____ therein. Further, he/she acknowledges that _____ cannot receive such inventions, discoveries, or information in confidence and will receive the disclosure thereof under the following conditions, and not otherwise.

1.   All rights and remedies of the undersigned and his assigns (and the principals of the undersigned, if any) arising from any disclosure made in connection with the above subject, by reference to materials listed herein, orally, in writing, or in any subsequent disclosure, to _____ or the use thereof by _____ shall be limited to those which may have been or in the future accorded to him/her, his/her assigns, or his/her principals by United States or foreign patents or copyrights.

2.   The undersigned hereby waives all other claims against _____ arising from any disclosure made by him/her or those authorized by him/her to _____ hereunder.

Signature: _____
Date: _____

*   Please list all materials relating to inventions, discoveries, and information, including cover letter, which you would like to have _____ evaluate. Use reverse side if necessary.

175

IX.3 Summary

Some companies refuse to accept or review ideas from nonemployees because: (1) The ideas submitted are rarely of value, and (2) this practice can result in substantial legal liability to the receiving company. Adhere to the following procedure if your company is willing to accept such ideas: When you first receive an outsider's idea <u>do</u> <u>not</u> study it, copy it, or discuss it with the submitter. Instead, forward the materials to your lawyer who will correspond with the submitter accordingly. Once your lawyer becomes involved, and the submitter signs a confidential disclosure waiver, you or others in your department are free to review the idea.

PART X: OUTSIDE CONSULTANTS

Sometimes it is desirable to use outside consultants to help with special tasks. These persons may be part of a major multi-employee corporation or may be individuals such as college professors and retired company employees.

Most often, when working with a major corporation the terms of consultation (e.g. the work to be done, ownership of proprietary information and inventions, obligations of secrecy, compensation, and legal status of the parties) are clearly defined via extensive negotiation followed by a substantial formal contract. However, when dealing with individuals there is a tendency to overlook these important points and treat the relationship more informally. This can result in the consultant (and perhaps your company) not having a clear understanding of his rights and obligations. You should develop as much legal definition when dealing with the small organization or individual as when dealing with a major corporation. All consulting arrangements should be discussed with your lawyers and consummated by a formal written and signed contract.

X.1 What To Discuss And Protect

Any consulting contract needs to cover both (a) the scope of the work to be done by the consultant and your company, and (b) at least the points of protection listed here. While we are not suggesting that you act as your company's lawyer, you will reduce the chance of misunderstanding if you discuss the following points with the consultant before the agreement is written and signed:

(1)   The consultant will protect your company's confidential information.

Your company's confidential business and technical information may be among its most valuable assets. Hence, you do not want to lose, or lessen the value of, such information by giving others the right to disclose or use confidential business or technical information.

(2)   The consultant will not provide technical assistance to your competitors while he's working for you.

(3)   The consultant will assign to your company the rights to any inventions, trade secrets, and patents which stem from the consultant's work.

(4)   The consultant will not knowingly infringe on the copyrights of others by using copyrighted materials in any work he prepares for your company; and the consultant will agree in writing to transfer to your company the copyright for any work he was specifically hired to prepare for your company.

(5)   The consultant understands that he is not your company's legal representative.

(6)   The consultant understands that he is not entitled to any benefits provided to your company's employees.

(7)   The consultant will not assign his rights, or delegate his duties, to another party.

Figure 14 is a sample contract designed to cover the above points while being fair to the consultant.

Figure 14: Sample Consulting Contract

Confidential Information

Consultant will hold in confidence and will not disclose to any third person any confidential information or trade secrets disclosed to Consultant by _____, its subsidiaries and/or affiliates during the course of his services hereunder. For purposes hereof, the words "confidential information" shall mean any and all information of _____, its subsidiaries and/or affiliates, regarding the products, customers, pricing, terms of sale, research and development or otherwise relating to the business of _____, its subsidiaries and/or affiliates, which is not generally available to third persons.

Noncompetition

During the Consulting Period, Consultant will not, without _____ prior written approval, establish or engage in any competitive business, or assist any other person to establish or engage in any competitive business, directly or through any enterprise or company in which Consultant is interested or acts as a partner, shareholder, director, officer, employee, or consultant. For purposes hereof, the words "competitive business" shall mean any business relating to the manufacture or sale of _____.

Inventions and Patents

Assignment -- Consultant agrees to assign to _____ without additional compensation, Consultant's entire right, title, and interest in and to any discoveries or inventions, whether or not patentable, which Consultant conceives or develops, individually or jointly, during or in connection with his performance or services hereunder and which relate to the production of _____.

Notice of Discoveries -- Consultant will report to _____ each discovery and/or invention described above promptly after Consultant conceives or discovers the same.

179

Cooperation -- Upon _____ request, Consultant will execute and deliver to _____ all papers and documents, and will perform all other acts, which _____ may deem necessary or appropriate in order to perfect the transfer to _____ of Consultant's right, title, and interest in and to each discovery and invention described above and to enable _____, at _____ election and at _____ expense, to file and obtain patents in _____ name under the laws of the United States of America and any foreign countries designated by _____ .

Copyrights

Neither Consultant nor any of Consultant's employees or independent contractors shall knowingly incorporate in any work prepared under this Agreement any copyrighted or proprietary material of another. Further, any work of authorship created under this Agreement shall constitute a "work made for hire," when so defined by the Copyright Act, and as to any work not so defined, Consultant hereby transfers to _____ any and all right, title, and interest Consultant may have in and to the copyright in such work.

Status and Authority

Independent Contractor -- Consultant will perform all services hereunder as an independent contractor. Nothing herein will be deemed to create an employee-employer or agent-principal relationship between Consultant and _____ .

Authority -- Nothing herein will be deemed to authorize Consultant to act as _____ agent or legal representative. Consultant hereby acknowledges that he is not authorized to act as _____ legal representative or otherwise.

Benefit Plans -- Consultant acknowledges that he will not be entitled to participate as an employee in or under any employee benefit plan of _____ or other employment rights or benefits available to or enjoyed by the employees of _____ .

Assignment

This Agreement will be deemed to require the performance of personal services by Consultant. Consultant will not assign any right, delegate any duty or otherwise transfer any interest hereunder without _____ prior written approval.

Notices

All notices and other communications required or permitted to be given hereunder, if in written form, will be deemed given two days after deposit in the U.S. mail, postage prepaid and addressed to the parties at their respective addresses set forth below (unless by written notice a different person or address shall have been designated.)

If to _____, to:

_____

_____

_____

_____

If to Consultant, to:

_____

_____

_____

_____

X.2 Summary

Any agreements of consultation, whether they be with a major corporation or an individual, should be documented via a signed formal contract. The contract should at least contain clauses on confidential information, noncompetition, inventions and patents, copyrights, authority, assignment of duties, and to whom notices are to be sent.

PART XI: TRADEMARKS

For engineering personnel, trademarks have less importance than any other subject in this book, since usually a company's marketing and legal departments have the primary responsibility for selecting, obtaining, and enforcing trademark rights. Oftentimes, however, in many industries a particular trademark is selected because the engineering department assigned that particular name or symbol to the engineering project long before the trademark was even considered. Here's how it usually works:

> The engineering personnel create a short, symbolic name for a product or project, and continually use that name in front of other company personnel and potential customers, and soon the company gravitates to using that name as the product's trademark.

In light of this way of life, it's best that engineering personnel have a general knowledge of what trademarks are all about so that they can better direct the naming of the engineering projects; that is, select names which have a chance of being viable trademarks. There is another (albeit less serious) reason for knowing about trademarks. Since trademarks are of interest to most consumers, trademark law makes for great cocktail party conversations!

XI.1 What's A Trademark?

> "A trademark is any word, name, symbol, or device (or any combination of these) that is adopted and used by a manufacturer or seller to identify his commercial goods or services in order to distinguish his goods or services from similar goods or services manufactured or sold by others."

Translated, this means that trademarks and service marks are used to let buyers know that the product or service carrying the mark is being provided by a particular company; and that the buyer can reasonably expect the quality of the product to be consistent with what the company usually provides. Thus, the real purpose of trademarks is to promote a company's products or services based on the company's reputation for quality.

Trademark laws are designed to eliminate consumer confusion. To this end, trademark law prevents competitors from using the same or a confusingly similar trademark on the same or similar goods which are sold in the same or related markets. Thus, a trademark owner can safely invest in advertising his product without fear that his advertisement dollars will benefit his competitors.

Some well-known trademarks are:

*   Coke® and Coca-Cola® for soft drinks (trademarks of the Coca-Cola Company, Atlanta, Georgia)

*   Seagrams® for whiskey and other liquors (a trademark of Joseph E. Seagram & Sons Inc., New York, New York)

*   Xerox® for copying and other business machines (a trademark of the Xerox Corporation, Stamford, Connecticut)

*   Ford® for automobiles (a trademark of the Ford Motor Company, Dearborn, Michigan)

*   IBM® for computers (a trademark of International Business Machines Corporation, Armonk, New York)

Usually, no special form is required for a trademark. For example, Coke® and Coca-Cola® are registered trademarks for soft drinks whether the letters are capitalized, small, or in script, block, or other form.

Federal law provides strong protection for trademarks. In the United States, trademark rights are based on how extensively and how long the mark is used in commerce; and the longer a trademark is used the stronger the trademark gets. This means that the more years your company uses a particular trademark, the more likely it is that your company has legal recourse against others who use the same or confusingly similar trademarks on similar products.

XI.2 "Trademarks" vs. "Trade Names"

Trademarks are often confused with trade names. As explained above, a trademark is a word, name, symbol, or device used by a manufacturer or merchant to identify his goods or services. Trade names, on the other hand, identify companies or vocations.

Specifically, trade names are used by:

(a)   manufacturers, industrialists, merchants, agriculturalists, and others to identify their businesses, vocations, or occupations; and

(b)   persons, firms, associations, companies, unions, and other organizations who lawfully adopt names or titles for their business entity.

Some famous trade names are E.I du Pont de Nemours and General Motors Corporation.

Oftentimes, the same identification is used for both the trademark and the trade name. For example, IBM is an acronym used to identify International Business Machines Corporation (a trade name) and the computers made by International Business Machines Corporation are sold as IBM® brand computers (a trademark). Hence, "IBM" is both a trade name and a trademark of International Business Machines Corporation.

XI.3 Why Register A Trademark And With Whom?

A company has three options with regard to registering a trademark in the United States:

(1)   They can register a trademark with one or more of the 50 states; or
(2)   They can obtain a federal registration; or
(3)   They can use a trademark without registering it with anyone.

If a trademark is registered in only one state, then that trademark is only protected within that one state; i.e. state registration must be done on a state-by-state basis. However, since the federal government has virtually absolute control over interstate commerce and such federal laws pre-empt any state laws, if a company can obtain federal registration for their trademark, their trademark is effective in all 50 states. In addition -- and as another

185

advantage of federal registration -- five years after obtaining federal registration the trademark cannot be contested by another company. For these reasons, most large U.S. companies obtain federal registration for their trademarks.

For a federal registration to be obtained and maintained a trademark must adhere to certain rules. Since state trademark laws vary from state to state, and more than likely your firm deals in interstate commerce and therefore obtains federal registrations, the remainder of our discussion on trademarks will deal only with federally registered trademarks.

XI.4 Federal Registration Requirements

Federal trademark registration gives the trademark owner exclusive, long-term rights to use the trademark throughout the states, territories, and possessions of the United States. However, federal trademark law contains rigid requirements which must be met before a trademark is entitled to federal registration.

Generally:

(1)   The trademark must be placed on the goods, containers, or tags or labels affixed to the product; and

(2)   The goods must have been sold or transported in interstate commerce.

Also, under federal law, the trademark can not be registered if the mark is:

(3)   Confusingly similar to the mark of another.

   For example, it is very doubtful that the Patent and Trademark Office would permit registration of "Serox" as a trademark for duplicating machines since "Serox" can be easily confused with the Xerox® mark for copy machines. A company can not obtain a federal registration on a trademark that is confusingly similar to another's trademark even though the other's trademark is not registered with the federal government.

(4)   Merely descriptive or deceptively misdescriptive of the goods.

        This is best explained by example. A trademark which is <u>merely descriptive</u> and is also a <u>generic description</u> would be "Copy Machine" for a copy machine. An example of a <u>deceptively misdescriptive</u> trademark would be "Bordeau" (an area of France) for a wine produced in California.

(5)   Primarily merely a surname.

        An example of a trademark which is primarily merely a surname would be "Jones" or "Smith" and probably could not be registered in connection with any product.

(6)   Primarily geographically descriptive or deceptively geographically misdescriptive.

        An example of a trademark which is primarily geographically descriptive would be "California" for wines made in California and an example of a deceptively geographically misdescriptive trademark could be "Bordeau" for wines made in California.

(7)   Comprised of "immoral, deceptive, or scandalous matter, or matter which may disparage or falsely suggest a connection with persons, living or dead, institutions, beliefs, or national symbols, or bring them into contempt or disrepute."

(8)   Comprised of the flag or coat of arms, or a simulation of the flag or coat of arms, of any country, state, or municipality.

(9)   Comprised of the name, portrait, or signature of a living person without his written consent or comprised of the name, signature, or portrait of a deceased U.S. president during the life of his widow except with her written consent.

As you read items (3) through (9) above some popular trademarks probably came to mind which seemed to violate the federal registration rules. You can rest assured that you're probably right -- the marks don't adhere to the rules. Most often, these marks were approved because the companies extensively used the mark long before they attempted to

187

obtain federal registration and, therefore, the mark was an established trademark and not adopted to deceive the consumers.

XI.5 Trademark Symbols

Because the use of the trademark gives the owner stronger legal rights, a trademark owner is encouraged to use his mark before it is registered; but a trademark owner cannot claim to have a registered trademark (i.e. indicated by an "R" in a circle) unless he has actually obtained a federal registration. Before obtaining federal registration, a trademark owner is entitled to provide notice of his claim to the trademark by placing a small "T.M." near the upper right hand end of the trademark (where owners of federally registered trademarks normally put the "R" in the circle). He can also elaborate in some other way that he is claiming trademark rights in the mark he has chosen. However, neither of these approaches (placing a "T.M." or otherwise stating that he is claiming trademark rights) will improve his legal rights in the trademark, but serve only to notify others that he considers himself to be the owner of the trademark. In other words, putting the "T.M." on the goods may discourage a potential competitor from adopting the same or a similar trademark.

XI.6 The Life Of A Trademark

A federal trademark registration remains in force for 20 years and can be renewed indefinitely for additional 20-year periods. However, the life of a federally registered trademark can be cut short under the following conditions:

(1)   *Federal trademark rights can be lost if the trademark becomes the common descriptive name of the goods.*

A trademark owner is in danger of losing his legal trademark rights if his trademark begins to lose its identification with the brand of product or service to which it was specifically assigned, and becomes a term which is used to describe all of those same kinds of products or services. Or to put it another way, if a trademark irreversibly changes from being an adjective to being a noun then all legal rights are lost forever.

188

Over the years, there have been companies that allowed their trademarks to become nouns and so lost their trademark rights. For example, the words aspirin, cellophane, linoleum, and refrigerator were once enforceable, valuable trademarks. "Aspirin" was a trademark for a salicylic acid pain killer, "linoleum" was a trademark for a sheet floor covering, "cellophane" was a trademark for clear plastic wrap, and "refrigerator" was a trademark for electric food cooling machines. As these products became popular, their trademarks became household words. All salicylic acid painkillers became known simply as aspirin, all plastic wraps became known as cellophane, all sheet floor coverings were known as linoleum, and all food cooling machines became known as refrigerators -- and all of these trademarks were lost. Ironically, all of these trademarks became so popular that they could no longer be considered trademarks!

Today, one company which vigorously enforces its legal trademark rights is the Coca-Cola Company. You have probably been in the situation where you asked for a "Coke" at a refreshment stand only to be told that the stand serves "Pepsi" or another brand of cola. You were probably asked if Pepsi® or another brand of cola would be a suitable replacement for Coke®. The reason that the concessionaire asked you if Pepsi® would be a suitable substitute for Coke® is that the Coca-Cola Company, in keeping with its trademark rights, has not allowed even the smallest vendor to sell any other brand of cola when Coke® is requested. The Coca-Cola Company does not want -- and rightfully so -- all colas to become known as "Coke." Aggressive enforcement is necessary if the trademark owner is to fully preserve his exclusive legal right to use the trademark.

Note that the Coca-Cola Company does not have the exclusive right to use the word "cola" for soft drinks. This can be seen by trademarks of manufacturers of other brands of soft drinks such as Pepsi® Cola and RC® Cola (respectively, trademarks of the Pepsi Cola Company, Purchase, New York and the Royal Crown Cola Co., Columbus, Georgia). Therefore, if you merely ask for a "cola" the concessionaire can give you any brand of "cola."

(2)   *Federal trademark rights can be lost if the trademark is abandoned.*

This is a question of intent (i.e. did the owner intend to abandon the mark) on the part of the trademark owner, but not using the trademark for two years creates a presumption that the owner intended to abandon the trademark.

(3)   *Federal trademark rights can be lost if the registration was fraudulently obtained in the first place.*

(4)   *Federal trademark rights can be lost if the trademark is used by, or with the permission of, the owner to misrepresent the source of the goods.*

(5)   *Federal trademark rights can be lost if there is misuse of a registration by applying the registration notice (i.e. "R" in a circle) to goods when the mark is not actually registered or when the mark is used for goods which are not listed in the registration.*

The legal protection provided by a federal trademark registration extends only to the products listed in the trademark registration. For example, a company may have a registered trademark for the term XL-Lent brand food products. The registration then entitles the company to use the trademark with an "R" in the circle for any reference (e.g. labels, advertisements, etc.) to their food products. However, if the company uses the same trademark with an "R" in the circle for driveway cleaners they are misusing the federal trademark registration and are jeopardizing their federally protected legal rights in the trademark for food products. They may be free to use "XL-Lent" as a trademark for the driveway cleaner without using the "R" in a circle -- or can use it with a "T.M." if they wish -- but they cannot use the "R" in the circle with the mark until it is actually registered for the cleaner.

(6)   *Federal trademark rights can be lost if the trademark owner fails to object to the use of closely similar marks by others.*

If the ABC Company uses "ABC" as a trademark for widgets and another manufacturer comes out with a "KBC" widget, consumers may become confused by the similar sound of "ABC" and "KBC." Therefore, if the ABC Company fails to do something about enforcing their "ABC" trademark they are not only jeopardizing their legal right to prevent the second manufacturer from using "KBC" as as trademark for widgets, but they are also opening the door to allow other manufacturers to use "ABC" or other confusingly similar trademarks for widgits. This means that the ABC Company can lose their trademark rights if they don't object to the "KBC" trademark.

XI.7 When and Why You Need A Trademark Specialist

Most trademark matters require the services of a trademark specialist. Even though most patent lawyers have some trademark experience, it's usually best for a company to work with patent or other lawyers that specialize in trademark law.

Specifically, a good trademark specialist can:

(1)   Give overall advice on the proper use of any trademark; e.g. when it can be marked as registered, etc.

(2)   Have a search made to determine if your company's proposed mark is unique or "confusingly similar" to a trademark of another and provide guidance on whether or not the proposed mark can be registered with the federal government.

Such a search will minimize the chances of your company's chosen trademark conflicting with previously used marks, and help to avoid the expense of later changing the mark and/or destroying advertising or other written materials.

(3)   Handle all legal requirements and documents necessary for obtaining federal registration.

(4)   Be involved in any decision to challenge the use of a mark being used by another company.

(5)   Have a search made to determine if the trademarks used by others conflict with any of your company's existing trademarks.

It may be difficult for the lawyer to become aware of all the potentially conflicting trademarks. Therefore, to aid him in his search, everyone in your company should be on the lookout for marks which may conflict with your company's proposed trademark.

Use Figure 15 if you are asked to provide trademark information to your trademark specialist. In fact, you may want to give a copy of Figure 15 to your marketing department so that they can also properly and crisply convey such trademark information. The document does not have to be notarized, signed, or witnessed but the information provided should be accurate.

<u>Figure 15: Trademark Data</u>

1.   Identify the mark for which registration is desired:

_____

2.   Has a search been made to determine if the mark is confusingly similar to someone else's mark? Yes ___ No ___

3.   Identify the goods on which mark is (or will be) used:

_____

_____

_____

_____

4.   Do you expect to also use the mark on other goods? Yes ___ No ___

5.   If yes, specify goods and expected date of use:

_____

_____

_____

_____

6.   On what date did you first use the mark? _____

7.   For what purpose did you first use the mark?
        _____ Intrastate Shipment
        _____ Interstate Shipment
        _____ Advertising
        _____ Catalog
        _____ Sales Bulletin
        _____ Other -- Specify _____

8.   On what date did you ship in U.S. interstate commerce the first goods having the mark applied? _____

193

9.   How is the mark applied to the good, i.e. cast in housing, stenciled in indelible ink, on tag or label affixed to good, on shipping label, etc.?

_____

_____

Note: If the mark is on a tag or label, attach seven (7) specimens of the tag or label now in use. If the specimens are bulky (e.g. mark printed directly on large package or embossed, cast, etc.) attach seven (7) photographs showing how mark is applied. If a different label or marking system was previously used on the first interstate shipment, please specify, describe, or send a specimen.

XI.8 Summary

A trademark is a promotional word, name, symbol, or device which lets buyers know that they can expect that the product carrying the trademark will be of the same quality as other products carrying the same trademark. Trademarks identify goods or services and should not be confused with trade names, which identify companies or vocations.

A trademark need not be registered with any state or federal agency, but it's a good idea to obtain federal trademark registration for any goods that will be involved in interstate commerce. To qualify for such federal registration the trademark must be sold or transported in interstate commerce, must not violate certain restrictions, and must be placed on the goods (or its containers or tags). A trademark owner cannot claim to have a registered trademark unless he, in fact, has obtained a federal registration -- then and only then can the trademark carry the "R" in a circle. Before obtaining federal registration, the trademark can carry a small "T.M." Trademark rights can be lost (and are lost) for a variety of reasons. Almost always, trademark specialists are required when dealing with trademark matters.

APPENDIX A: **SOFTWARE PROTECTION**

Software involves a number of different intellectual property concepts, each of which may be entitled to different protection. This section on software protection will not only give you valuable information about the legal protection of software but will also serve an example of the various forms of intellectual property protection.

Every software product has at least three components: (1) the process, idea, or system upon which the program is based; (2) the program itself written in some programming language; and (3) the disk or other substrate upon which the program is written. The process may be protected under the patent law or even the trade secret law. On the other hand, the program itself can be protected by the copyright law but also may be protected as a trade secret, and the disk may be protected under the patent law.

The following describes how these three concepts (i.e. patent law, copyright law, and the laws of trade secrets) can be used to protect software.

I. Patent Law Applied to Software

Computer programs _per se_ are not patentable. However:

(1)   The disk or other substrate upon which a program is written may be patentable as a hardware item.

(2)   The software program itself may reduce an idea to practice and therefore it may be patentable as a process. (For a discussion on reducing an idea to practice, see Part V, Chapter 1.) The result of this process need not be, of itself, patentable. However, like any other invention, to be patentable the software program must be new, useful, and unobvious to one skilled in the particular subject matter.

Note: Processes (like machines, manufactures, or compositions of matter) are (a) not protected by the patent law if they are mere ideas which are not reduced to practice; (b) not protected by patent law if they are natural laws; (c) not protected by the patent law if they are forces and principles; and (d) not protected by the patent law if they are mental steps.

Most of the software produced today can not be protected by the patent laws. This is because much of the creativity upon which computer software is based consists of mathematical formulae; and the courts have specifically held that an abstract process embodied in computer software which is a mathematical method of converting binary numbers to digital numbers is not patentable (*Gottschalk vs. Benson, 409 U.S. 63, 1972*). It's important to realize, however, that the Supreme Court specifically stated that it was not saying that all computer programs were nonpatentable. In fact, in a later decision, the Supreme Court stated that the use of a mathematical equation and computer program in a process did not, in and of itself, render that process unpatentable (*Diamond vs. Diehl, 450 U.S. 175, 1981*).

II. Copyright Law Applied to Software

Ordinarily, copyright law entitles the original creator to several exclusive rights to his work, the most important of which is the right to prevent others from reproducing his work without his permission beyond the limits of "fair use." These exclusive rights generally remain in force for the lifetime of the author plus fifty years after his death. However, computer software is subject to exceptions to these otherwise comprehensive rights.

While computer programs have been accepted for copyright registration since 1964, the 1976 Copyright Law provided a statutory basis for including computer programs. Sections 101 and 117, as amended, provide that computer programs are copyrightable and that the Act applies to all computer uses of copyrighted programs; but that *owners of COPIES of copyrighted programs are allowed to reproduce these copies as necessary for effective use without incurring liability for infringement.*

In addition to such legal copying, illegal copying of computer programs is commonplace. Since copyrights are not self-enforcing the owner of a copyrighted software program must find the infringing copies and their producers, and bring appropriate legal action. Because the equipment required for copying is readily available (e.g. usually any mini-computer) finding the infringer is often a difficult

-- if not an impossible -- task. In addition to the difficulties faced in identifying the offending parties, copyright infringement actions are oftentimes consuming and complex, with adequate recovery of damages not a certainty. While many copyright infringement cases are successfully pursued, a large number goes undiscovered, while in still others no action is brought.

> For more information on copyrighting software -- including what steps to take -- see Figure 16, which is an information circular recently distributed by the Copyright Office. Note that the same Form TX is used which is included in Part VI of this book.

In summary, while the copyright law does in fact apply to computer software programs, a specific ruling in that law allows for legal copying. Further, as a practical matter, it's very difficult to bring legal action against persons who copy the program illegally.

III. Trade Secret Law Applied to Software

Software could be protected under trade secret concepts if the software were closely held or not capable of being deciphered. However, if a legitimately obtained copy of a computer program were deciphered the author would have no protection against a party who decided to sell it for profit.

DEFINITION

"A 'computer program' is a set of statements or instructions to be used directly or indirectly in a computer in order to bring about a certain result."

WHAT TO SEND

- A Completed Form TX
- A $10.00 Non-refundable Filing Fee Payable to the Register of Copyrights
- One Copy of Identifying Material (See Below)

EXTENT OF COPYRIGHT PROTECTION

Copyright protection extends to the literary or textual expression contained in the computer program. Copyright protection is not available for ideas, program logic, algorithms, systems, methods, concepts, or layouts.

DESCRIBING BASIS OF CLAIM ON FORM TX

- Space 2. In the "Author of" space identify the copyrightable authorship in the computer program for which registration is sought; for example, AUTHOR OF "Text of computer program," "Text of user's manual and computer program text," etc. (Do not include in the claim any reference to design, physical form, or hardware.)
- Space 6. Complete this space only if the computer program contains a substantial amount of previously published, registered, or public domain material (for example, subroutines, modules, or textual material).

DEPOSIT REQUIREMENTS

For published or unpublished computer programs, one copy of identifying portions of the program, (first 25 and last 25 pages), reproduced in a form visually perceptible without the aid of a machine or device, either on paper or in micro-

form, together with the page or equivalent unit containing the copyright notice, if any.

The Copyright Office believes that the best representation of the authorship in a computer program is a listing of the program in source code.

Where the applicant is unable or unwilling to deposit a source code listing, registration will proceed under our RULE OF DOUBT policy upon receipt of written assurance from the applicant that the work as deposited in object code contains copyrightable authorship.

If a published user's manual (or other printed documentation) accompanies the computer program, deposit two copies of the user's manual along with one copy of the identifying portion of the program.

DEPOSIT REQUIREMENTS
Section 202.20(c) (vii), 37 C.F.R.

(vii) *Machine-readable works.* In cases where an unpublished literary work is fixed or a published literary work is published only in the form of machine-readable copies (such as magnetic tapes or disks, punched cards, or the like) from which the work cannot ordinarily be perceived except with the aid of a machine or device,[4] the deposit shall consist of:

(A) for published or unpublished computer programs, one copy of identifying portions of the program, reproduced in a form visually perceptible without the aid of a machine or device, either on paper or in microform. For these purposes, "identifying portions" shall mean either the first and last 25 pages or equivalent units of the program if reproduced on paper, or at least the first and last 25 pages or equivalent units of the program if reproduced in microform, together with the page or equivalent unit containing the copyright notice, if any.

---

[4] Works published in a form requiring the use of a machine or device for purposes of optical enlargement (such as film, filmstrips, slide films, and works published in any variety of microform) and works published in visually perceptible form but used in connection with optical scanning devices, are not within this category.

Figure 16: Copyright Registration For Computer Programs

LOCATION OF COPYRIGHT NOTICE
Section 201.20(g), 37 C.F.R.

(g) WORKS REPRODUCED IN MACHINE-READABLE COPIES.

For works reproduced in machine-readable copies (such as magnetic tapes or disks, punched cards, or the like, from which the work cannot ordinarily be visually perceived except with the aid of a machine or device,[1] each of the following constitute examples of acceptable methods of affixation and position of notice:

(1) A notice embodied in the copies in machine-readable form in such a manner that on visually perceptible print-outs it appears either with or near the title, or at the end of the work;

(2) A notice that is displayed at the user's terminal at sign on;

(3) A notice that is continuously on terminal display; or

(4) A legible notice reproduced durably, so as to withstand normal use, on a gummed or other label securely affixed to the copies or to a box, reel, cartridge, cassette, or other container used as a permanent receptacle for the copies.

FORM OF COPYRIGHT NOTICE

Form of Notice for Visually Perceptible Copies

The notice for visually perceptible copies should contain all of the following three elements:

1. *The symbol* © (the letter C in a circle), or the word "Copyright," or the abbreviation "Copr."

2. *The year of first publication* of the work. In the case of compilations of derivative works incorporating previously published material, the year date of first publication of the compilation or derivative work is sufficient.

3. *The name of the owner of copyright* in the work, or an abbreviation by which the name can be recognized, or a generally known alternative designation of the owner.

Example: © 1983 John Doe

FURTHER QUESTIONS:

If you have general information questions and wish to talk to an information specialist, call 202-287-8700.

TO ORDER FORMS:

Write to Information and Publications Section, LM-455, Copyright Office, Library of Congress, Washington, D.C. 20559 or call 202-287-9100, the Forms Hotline.

Please note that a copyright registration is effective on the date of receipt in the Copyright Office of all the required elements in acceptable form, regardless of the length of time it takes thereafter to process the application and mail the certificate of registration. The length of time required by the Copyright Office to process an application varies from time to time, depending on the amount of material received and the personnel available to handle it. It must also be kept in mind that it may take a number of days for mailed material to reach the Copyright Office and for the certificate of registration to reach the recipient after being mailed by the Copyright Office.

If you are filing an application for copyright registration in the Copyright Office, you will not receive an acknowledgement that your application has been received (the Office receives more than 500,000 applications annually), but you can expect:

- A letter or telephone call from a copyright examiner if further information is needed;
- A certificate of registration to indicate the work has been registered, or if the application cannot be accepted, a letter explaining why it has been rejected.

You may not receive either of these until 90 days have passed.

If you want to know when the Copyright Office received your material, you should send it via registered or certified mail and request a return receipt.

Copyright Office • Library of Congress • Washington, D.C. 20559

✩ U.S. Government Printing Office: 1983—381-279: 572

Figure 16: Copyright Registration For Computer Programs

The Economics of Software Protection

The economics of software development, combined with the unique nature of the software/hardware interface, make the issue of software protection nettlesome at best.

Packaged software, which is generally sold on magnetic discs, is often expensive -- ranging upwards from $100 - $500 for personal computer programs. However, the discs themselves (which are the only tangible part of the software) usually sell for $1 - $5 each. This price spread alone encourages copying rather than purchasing packaged programs. In addition, the fact that almost anyone using the software has the capability at hand to duplicate the program makes the problem even more difficult for the manufacturer of the original software program.

Many unauthorized copies of computer software occur when rightful owners of a program reproduce it for multiple use to avoid purchasing additional copies. Another type of infringement occurs when a program is copied from an original for distribution to a third party, whether for sale or free. All of these actions take money out of the creator's pocket.

The solution to the problem of software protection for the manufacturer may lie in technology rather than a legal "fix." Software producers have tried various codes to frustrate would be copiers; however, a determined and skillful computer operator is usually successful in eventually breaking the code and copying the software. In fact, for just a few dollars one can purchase "code breaking" software which makes copying quick and easy. One promising development in the area of hardware protection schemes (which would merely reduce the ease of copying) is the use of weak bits within the program itself. These weak bits (bits read sometimes as an 0 and sometimes as a 1) would be intentionally included in a program such that they would not be able to be copied, thus rendering any copy inoperable. In addition, the cells could be designed to cease functioning after a given number of uses. Therefore, a purchaser would, in effect, buy a specific number of uses of the software package. Despite these developments, however, very few people believe technical protection schemes to be a panacea or ultimately unbreakable.

APPENDIX B: HOW TO GET THE PUBLICATIONS REFERRED TO IN THIS BOOK

Below is a list of all of the publications recommended in this guide and where they appear in the guide (i.e. part and chapter number). Except where indicated, they can be obtained from the:

Superintendent of Documents
U.S. Government Printing Office
Washington, DC 20402

(1)   *Attorneys and Agents Registered to Practice Before the U.S. Patent and Trademark Office* (Part II, "About Lawyers")

A directory of the more than 12,000 patent attorneys and agents registered to practice before the Patent and Trademark Office. The directory, which is arranged by state and city, sells for $17.00.

(2)   *Manual of Patent Office Classification* (Part III, "Overview Of United States Patents," and Part V, Chapter 10, "How And Where To Find Patent Information")

A publication which identifies the technical categories and subcategories into which the Patent and Trademark Office classifies patents. The manual sells for $77.00.

(3)   *Index to Manual of Patent Office Classification* (Part III, "Overview Of United States Patents," and Part V, Chapter 10, "How And Where To Find Patent Information")

A corresponding publication to *Manual of Patent Office Classification*. The cost of the index is $9.00.

(4)   *The Rules of Practice in Patent Cases* (Part V, Chapter 4, "What You Need To Know About Applying For A Patent")

A guide which outlines the complete requirements, rules, and regulations for patent application, including any drawing accompanying a patent application; specifying such details as sheet size, type of paper to be used, and margins. This guide can be obtained from:

> Rules Service Co.
> 7658 Standish Place, Suite 106
> Rockville, MD 20855
> (301) 424-9402

(5)   *Manual of Patent Office Examination Procedures* (Part V, Chapter 6, "What Happens After The Application Is Filed")

The complete procedure used by the Patent and Trademark Office in examining and evaluating a patent application; including evaluation for obviousness, the prior art search, the application's compliance with legal requirements, and review of the claims. This manual costs $70.00.

(6)   The *Official Gazette of the U.S. Patent and Trademark Office* (Part V, Chapter 10, "How And Where To Find Patent Information")

A weekly publication by the Patent and Trademark Office which lists each patent issued for that week, including patent number, inventor, assignee, and drawing. The *Official Gazette* also contains information on design patents and trademarks, and changes in patent laws and procedures. It is available by subscription from the U.S. Government Printing Office in Washington, D.C. The subscription price is $270.00 per year for subscribers living in the U.S. and $380.00 per year for subscribers living in a foreign country.

(7)   *Index of Inventors and Assignees* (Part V, Chapter 10, "How And Where To Find Patent Information")

A listing of all inventors and assignees, published annually by the Patent and Trademark Office.

(8)   The *Federal Register* (Part V, Chapter 11, "Patent Costs Itemized")

Publication which will allow you to find more information on patent fees. It can be obtained from:

> Commissioner of Patents and Trademarks
> Washington, DC 20231
> (703) 557-1610

(9)   *Code of Federal Regulations*, Circular R96.201.20 (Part VI, "Copyrights")

This circular contains regulations concerning the proper form and position of the copyright notice. This, and any other copyright information, can be obtained from:

> Copyright Office
> LM 455
> Library of Congress
> Washington, D.C. 20559

(10)   *Submitting An Idea* (Part IX, "Outsiders' Ideas")

An American Bar Association Booklet which explains the potential legal liabilities of accepting outside disclosures. The booklet can be obtained from:

> American Bar Association
> American Bar Center
> 750 North Lake Shore Drive
> Chicago, IL 60611

APPENDIX C: LIBRARIES HAVING U.S. PATENTS

The following libraries receive current issues of U.S. patents and maintain collections of earlier issued patents. The scope of these collections varies from library to library.

State	Name of Library	Phone
Alabama	Auburn University Libraries	(205) 826-4500, ext. 21
	Birmingham Public Library	(205) 226-3680
Alaska	Anchorage Municipal Libraries	(907) 264-4481
Arizona	Tempe: Noble Library, Arizona State University	(602) 965-7609
Arkansas	Little Rock: Arkansas State Library	(501) 371-2090
California	Los Angeles Public Library	(213) 612-3273
	Sacramento: California State Library	(916) 322-4572
	San Diego Public Library	(619) 236-5813
	Sunnyvale: Patent Information Clearinghouse	(408) 730-7290
Colorado	Denver Public Library	(303) 571-2122
Delaware	Newark: University Of Delaware Library	(302) 451-2965
Florida	Fort Lauderdale: Broward County Main Library	(305) 357-7444
	Miami-Dade Public Library	(305) 375-2665
Georgia	Atlanta: Price Gilbert Memorial Library, Georgia Institute of Technology	(404) 894-4508
Idaho	Moscow: University of Idaho Library	(208) 885-6235
Illinois	Chicago Public Library	(312) 269-2865
	Springfield: Illinois State Library	(217) 782-5430
Indiana	Indianapolis-Marion County Public Library	(317) 269-1741
Louisiana	Baton Rouge: Troy H. Middleton Library, Louisiana State University	(504) 388-2570
Maryland	College Park: Engineering and Physical Sciences Library, University of Maryland	(301) 454-3037
Mass.	Amherst: Physical Sciences Library University of Massachusetts	(413) 545-1370
	Boston Public Library	(617) 536-5400, ext. 265
Michigan	Ann Arbor: Engineering Transportation Library, University of Michigan	(313) 764-7494
	Detroit Public Library	(313) 833-1450

Minnesota	Minneapolis Public Library & Information Ctr.	(612) 372-6570
Missouri	Kansas City: Linda Hall Library	(816) 363-4600
	St. Louis Public Library	(314) 241-2288, ext. 390
Montana	Butte: Montana College of Mineral Science and	
	Technology Library	(406) 496-4284
Nebraska	Lincoln: University Of Nebraska-Lincoln,	
	Engineering Library	(402) 472-3411
Nevada	Reno: University of Nevada Library	(702) 784-6579
New Hamp.	Durham: University of New Hampshire	
	Library	(603) 862-1777
New Jersey	Newark Public Library	(201) 733-7815
New Mexico	Albuquerque: University of New Mexico	
	Library	(505) 277-5441
New York	Albany: New York State Library	(518) 474-7040
	Buffalo and Erie County Public Library	(716) 856-7525, ext. 267
	New York Public Library (The Research	
	Libraries)	(212) 714-8529
N. Carolina	Raleigh: D.H. Hill Library	
	N.C. State University	(919) 737-3280
Ohio	Cincinnati & Hamilton County Public Library	(513) 369-6936
	Cleveland Public Library	(216) 623-2870
	Columbus: Ohio State University Libraries	(614) 422-6286
	Toledo/Lucas County Public Library	(419) 255-7055, ext. 212
Oklahoma	Stillwater: Oklahoma State University Library	(405) 624-6546
Oregon	Salem: Oregon State Library	(503) 378-4239
Pennsylvania	Cambridge Springs: Alliance College Library	(814) 398-2098
	Philadelphia: Franklin Institute Library	(215) 448-1227
	Philadelphia: The Free Library	(215) 686-5330
	Pittsburgh: Carnegie Library of Pittsburgh	(412) 622-3138
	University Park: Pattee Library, Pennsylvania	
	State University	(814) 865-4861
Rhode Island	Providence Public Library	(401) 521-8726
S. Carolina	Charleston: Medical University of South	
	Carolina Library	(803) 792-2371

Tennessee	Memphis & Shelby County Public Library and Information Center	(901) 725-8876
	Nashville: Vanderbilt University Library	(615) 322-2775
Texas	Austin: McKinney Engineering Library, University of Texas	(512) 471-1610
	College Station: Sterling C. Evans Library, Texas A & M University	(409) 845-2551
	Dallas Public Library	(214) 749-4176
	Houston: The Fondren Library, Rice University	(713) 527-8101, ext. 2587
Utah	Salt Lake City: Marriott Library University of Utah	(801) 581-8394
Virginia	Richmond: Virginia Commonwealth University Library	(804) 257-1104
Washington	Seattle: Engineering Library University of Washington	(206) 543-0740
Wisconsin	Madison: Kurt F. Wendt Engineering Library, University of Wisconsin	(608) 262-6845
	Milwaukee Public Library	(414) 278-3247

APPENDIX D: **THE UNITED STATES CODE ON THE**
                PATENTABILITY OF INVENTIONS

Here are excerpts from Title 35 of the United States Code, which contain the basic legal requirements concerning patents. We have not included all patent legal requirements but only those which an engineering person may want to refer to in order to complement the contents of this book.

Patentability of Inventions

S100. Definitions
When used in this title unless the context otherwise indicates --

(a)   The term "invention" means invention or discovery.

(b)   The term "process" means process, art or method, and includes a new use of a known process, machine, manufacture, composition of matter, or material.

(c)   The terms "United States" and "this country" mean the United States of America, its territories and possessions.

(d)   The word "patentee" includes not only the patentee to whom the patent was issued but also the successors in title to the patentee.

S101. Inventions patentable
Whoever invents or discovers any new and useful process, machine, manufacture, or composition of matter, or any new and useful improvement thereof, may obtain a patent therefor, subject to the conditions and requirements of this title.

S102. Conditions for patentability; novelty and loss of right to patent
A person shall be entitled to a patent unless --

(a) the invention was known or used by others in this country, or patented or described in a printed publication in this or a foreign country, before the invention thereof by the applicant for patent, or

(b) the invention was patented or described in a printed publication in this or a foreign country or in public use or on sale in this country, more than one year prior to the date of the application for patent in the United States, or

(c) he has abandoned the invention, or

(d) the invention was first patented or caused to be patented, or was the subject of an inventor's certificate, by the applicant or his legal representatives or assigns in a foreign country prior to the date of the application for patent in this country on an application for patent or inventor's certificate filed more than twelve months before the filing of the application in the United States, or

(e) the invention was described in a patent granted on an application for patent by another filed in the United States before the invention thereof by the applicant for patent, or

211

(f) he did not himself invent the subject matter sought to be patented, or

(g) before the applicant's invention thereof the invention was made in this country by another who had not abandoned, suppressed, or concealed it. In determining priority of the invention there shall be considered not only the respective date of conception and reduction to practice of the invention, but also the reasonable diligence of one who was first to conceive and last to reduce to practice, from a time prior to conception by the other.

S103. Conditions for patentability; non-obvious subject matter

A patent may not be obtained though the invention is not identically disclosed or described as set forth in section 102 of this title, if the differences between the subject matter sought to be patented and the prior art are such that the subject matter as a whole would have been obvious at the time the invention was made to a person having ordinary skill in the art to which said subject matter pertains. Patentability shall not be negatived by the manner in which the invention was made. Subject matter developed by another person, which qualifies as prior art only under subsection (f) or (g) of section 102 of this title, shall not preclude patentability under this section where the subject matter and the claimed invention were, at the same time the invention was made, owned by the same person or subject to an obligation of assignment to the same person.

Application for Patent

S111. Application for patent

Application for patent shall be made, or authorized to be made, by the inventor, except as otherwise provided in this title, in writing to the Commissioner. Such application shall include (1) a specification as prescribed by section 112 of this title; (2) a drawing as prescribed by section 113 of this title; and (3) an oath by the applicant as prescribed by section 115 of this title. The application must be accompanied by the fee required by law. The fee and oath may be submitted after the specification and any required drawing are submitted, within such period and under such conditions, including the payment of a surcharge, as may be prescribed by the Commissioner. Upon failure to submit the fee and oath within such prescribed period, the application shall be regarded as abandoned, unless it is shown to the satisfaction of the Commissioner that the delay in submitting the fee and oath was unavoidable. The filing date of an application shall be the date on which the specification and any required drawing are received in the Patent and Trademark Office.

S112. Specification

The specification shall contain a written description of the invention, and of the manner and process of making and using it, in such full, clear, concise, and exact terms as to enable any person skilled in the art to which it pertains, or with which it is most nearly connected, to make and use the same, and shall set forth the best mode contemplated by the inventor for carrying out his invention.

The specification shall conclude with one or more claims particularly pointing out and distinctly claiming the subject matter which the applicant regards as his invention. A claim may be written in independent or dependent form, and if in dependent form it shall be construed to include all the limitations of the claim incorporated by reference into the dependent claim.

An element in a claim for a combination may be expressed as a means or step for performing a specified function without the recital of structure, material, or acts in support thereof, and such claim shall be construed to cover the corresponding structure, material, or acts described in the specification and equivalents thereof.

S113. Drawings

The applicant shall furnish a drawing where necessary for the understanding of the subject matter sought to be patented. When the nature of such subject matter admits of illustration by a drawing and the applicant has not furnished such a drawing, the Commissioner may require its submission within a time period of not less than two months from the sending of a notice thereof. Drawings submitted after the filing date of the application may not be used (i) to overcome any insufficiency of the specification due to lack of an enabling disclosure or otherwise inadequate disclosure therein, or (ii) to supplement the original disclosure thereof for the purpose of interpretation of the scope of any claim.

S115. Oath of applicant

The applicant shall make oath that he believes himself to be the original and first inventor of the process, machine, manufacture, or composition of matter, or improvement thereof, for which he solicits a patent: and shall state of what country he is a citizen. Such oath may be made before any person within the United States authorized by law to administer oaths, or, when made in a foreign country, before any diplomatic or consular officer of the United States authorized to administer oaths, or before any officer having an official seal and authorized to administer oaths in the foreign country in which the applicant may be, whose authority is proved by certificate of a diplomatic or consular officer of the United States, or apostille of an official designated by a foreign country which, by treaty or convention, accords like effect to apostilles of designated officials in the United States. Such oath is valid if it complies with the laws of the state or country where made. When the application is made as provided in this title by a person other than the inventor, the oath may be so varied in form that it can be made by him.

S116. Inventors

When an invention is made by two or more persons jointly, they shall apply for patent jointly and each make the required oath, except as otherwise provided in this title. Inventors may apply for a patent jointly even though (1) they did not physically work together or at the same time, (2) each did not make the same type or amount of contribution, or (3) each did not make a contribution to the subject matter of every claim of the patent.

If a joint inventor refuses to join in an application for patent or cannot be found or reached after diligent effort, the application may be made by the other inventor on behalf of himself and the omitted inventor. The Commissioner, on proof of the pertinent facts and after such notice to the omitted inventor as he prescribes, may grant a patent to the inventor making the application, subject to the same rights which the omitted inventor would have had if he had been joined. The omitted inventor may subsequently join in the application.

Whenever through error a person is named in an application for patent as the inventor, or through an error an inventor is not named in an application, and such error arose without any deceptive intention on his part, the Commissioner may permit the application to be amended accordingly, under such terms as he prescribes.

S117. Death or incapacity of inventor

Legal representatives of deceased inventors and of those under legal incapacity may make application for patent upon compliance with the requirements and on the same terms and conditions applicable to the inventor.

S118. Filing by other than inventor
Whenever an inventor refuses to execute an application for patent, or cannot be found or reached after diligent effort, a person to whom the inventor has assigned or agreed in writing to assign the invention or who otherwise shows sufficient proprietary interest in the matter justifying such action, may make application for patent on behalf of and as agent for the inventor on proof of the pertinent facts and a showing that such action is necessary to preserve the rights of the parties or to prevent irreparable damage; and the Commissioner may grant a patent to such inventor upon such notice to him as the Commissioner deems sufficient, and on compliance with such regulations as he prescribes.

S119. Benefit of earlier filing date in foreign country; right of priority
An application for patent for an invention filed in this country by any person who has, or whose legal representatives or assigns have, previously regularly filed an application for a patent for the same invention in a foreign country which affords similar privileges in the case of applications filed in the United States or to citizens of the United States, shall have the same effect as the same application would have if filed in this country on the date on which the application for patent for the same invention was first filed in such foreign country, if the application in this country is filed within twelve months from the earliest date on which such foreign application was filed; but no patent shall be granted on any application for patent for an invention which had been patented or described in a printed publication in any country more than one year before the date of the actual filing of the application in this country, or which had been in public use or on sale in this country more than one year prior to such filing.

No application for patent shall be entitled to this right of priority unless a claim therefor and a certified copy of the original foreign application, specification and drawings upon which it is based are filed in the Patent and Trademark Office before the patent is granted, or at such time during the pendency of the application as required by the Commissioner not earlier than six months after the filing of the application in this country. Such certification shall be made by the patent office of the foreign country in which filed and show the date of the application and of the filing of the specification and other papers. The Commissioner may require a translation of the papers filed if not in the English language and such other information as he deems necessary.

In like manner and subject to the same conditions and requirements, the right provided in this section may be based upon a subsequent regularly filed application in the same foreign country instead of the first filed foreign application, provided that any foreign application filed prior to such subsequent application has been withdrawn, abandoned, or otherwise disposed of, without having been laid open to public inspection and without leaving any rights outstanding, and has not served, nor thereafter shall serve, as a basis for claiming a right of priority.

Applications for inventors' certificates filed in a foreign country in which applicants have a right to apply, at their discretion, either for a patent or for an inventor's certificate shall be treated in this country in the same manner and have the same effect for purpose of the right of priority under this section as applications for patents, subject to the same conditions and requirements of this section as apply to applications for patents, provided such applicants are entitled to the benefits of the Stockholm Revision of of the Paris Convention at the time of such filing.

214

S120. Benefit of earlier filing date in the United States
An application for patent for an invention disclosed in the manner provided by the first paragraph of section 112 of this title in an application previously filed in the United States, or as provided by section 363 of this title, which is filed by an inventor or inventors named in the previously filed application shall have the same effect, as to such invention, as though filed on the date of the prior application, if filed before the patenting or abandonment of or termination of proceedings on the first application or on an application similarly entitled to the benefit of the filing date of the first application and if it contains or is amended to contain a specific reference to the earlier filed application.

S122. Confidential status of applications
Applications for patents shall be kept in confidence by the Patent and Trademark Office and no information concerning the same given without authority of the applicant or owner unless necessary to carry out the provisions of any Act of Congress or in such special circumstances as may be determined by the Commissioner.

Examination of Application

S131. Examination of application
The Commissioner shall cause an examination to be made of the application and the alleged new invention; and if on such examination it appears that the applicant is entitled to a patent under the law, the Commissioner shall issue a patent therefor.

S132. Notice of rejection; reexamination
Whenever, on examination, any claim for a patent is rejected, or any objection or requirement made, the Commissioner shall notify the applicant thereof, stating the reasons for such rejection, or objection or requirement, together with such information and references as may be useful in judging of the propriety of continuing the prosecution of his application; and if after receiving such notice, the applicant persists in his claim for a patent, with or without amendment, the application shall be reexamined. No amendment shall introduce new matter into the disclosure of the invention.

S133. Time for prosecuting application
Upon failure of the applicant to prosecute the application within six months after any action therein, of which notice has been given or mailed to the applicant, or within such shorter time, not less than thirty days, as fixed by the Commissioner in such action, the application shall be regarded as abandoned by the parties thereto, unless it be shown to the satisfaction of the Commissioner that such delay was unavoidable.

S134. Appeal to the Board of Appeals
An applicant for a patent, any of whose claims has been twice rejected, may appear from the decision of the primary examiner to the Board of Appeal, having once paid the fee for such appeal.

S135. Interferences
(a) Whenever an application is made for a patent which, in the opinion of the Commissioner, would interfere with any pending application, or with any unexpired patent, an interference may be declared and the Commissioner shall give notice of such declaration to the applicants, or applicant and patentee, as the case may be. The Board of Patent Appeals and Interferences shall determine questions of priority of the inventions and may determine questions of patentability. Any final decision, if adverse to the claim of an applicant, shall constitute the final refusal by the Patent and Trademark Office of

the claims involved, and the Commissioner may issue a patent to the applicant who is adjudged the prior inventor. A final judgment adverse to a patentee from which no appeal or other review has been or can be taken or had shall constitute cancellation of the claims involved in the patent, and notice of such cancellation shall be endorsed on copies of the patent distributed after such cancellation by the Patent and Trademark Office.

(b) A claim which is the same as, or for the same or substantially the same subject matter as, a claim of an issued patent may not be made in any application unless such a claim is made prior to one year from the date on which the patent was granted.

(c) Any agreement or understanding between parties to an interference, including any collateral agreements referred to therein, made in connection with or in contemplation of the termination of the interference, shall be in writing and a true copy thereof filed in the Patent and Trademark Office before the termination of the interference as between the said parties to the agreement or understanding. If any party filing the same so requests, the copy shall be kept separate from the file of the interference, and made available only to Government agencies on written request, or to any person on a showing of good cause. Failure to file the copy of such agreement or understanding shall render permanently unenforceable such agreement or understanding and any patent of such parties involved in the interference or any patent subsequently issued on any application of such parties so involved. The Commissioner may, however, on a showing of good cause for failure to file within the time prescribed, permit the filing of the agreement or understanding during the six-month period subsequent to the termination of the interference as between the parties to the agreement or understanding.

The Commissioner shall give notice to the parties or their attorneys of record, a reasonable time prior to said termination, of the filing requirement of this section. If the Commissioner gives such notice at a later time, irrespective of the right to file such agreement or understanding within the six-month period on a showing of good cause, the parties may file such agreement or understanding within sixty days of the receipt of such notice.

Any discretionary action of the Commissioner under this subsection shall be reviewable under section 10 of the Administrative Procedure Act.

(d) Parties to a patent interference, within such time as may be specified by the Commissioner by regulation, may determine such contest or any aspect thereof by arbitration. Such arbitration shall be governed by the provisions of title 9 to the extent such title is not inconsistent with this section. The parties shall give notice of any arbitration award to the Commissioner, and such award shall, as between the parties to the arbitration, be dispositive of the issues to which it relates. The arbitration award shall be unenforceable until such notice is given. Nothing in this subsection shall preclude the Commissioner from determining patentability of the invention involved in the interference.

Issue of Patent

S154. <u>Contents and term of patent</u>
Every patent shall contain a short title of the invention and a grant to the patentee, his heirs or assigns, for the term of seventeen years, subject to the payment of issue fees as provided for in this title, of the right to <u>exclude others from</u> making, using, or selling the invention throughout the United States, referring to the specification for the particulars

thereof. A copy of the specification and drawings shall be annexed to the patent and be a part thereof.

Secrecy of Certain Inventions and Filing Applications in Foreign Countries

S181. Secrecy of certain inventions and withholding of patent
Whenever publication or disclosure by the grant of a patent on an invention in which the Government has a property interest might, in the opinion of the head of the interested Government agency, be detrimental to the national security, the Commissioner upon being so notified shall order that the invention be kept secret and shall withhold the grant of a patent therefor under the conditions set forth hereinafter.

Whenever the publication or disclosure of an invention by the granting of a patent, in which the Government does not have a property interest, might, in the opinion of the Commissioner, be detrimental to the national security, he shall make the application for patent in which such invention is disclosed available for inspection to the Atomic Energy Commission, the Secretary of Defense, and the chief officer of any other department or agency of the Government designated by the President as a defense agency of the United States.

Each individual to whom the application is disclosed shall sign a dated acknowledgment thereof, which acknowledgment shall be entered in the file of the application. If, in the opinion of the Atomic Energy Commission, the Secretary of a Defense Department, or the chief officer of another department or agency so designated, the publication or disclosure of the invention by the granting of a patent therefor would be detrimental to the national security, the Atomic Energy Commission, the Secretary of a Defense Department, or such other chief officer shall notify the Commissioner and the Commissioner shall order that the invention be kept secret and shall withhold the grant of a patent for such period as the national interest requires, and notify the applicant thereof. Upon proper showing by the head of the department or agency who caused the secrecy order to be issued that the examination of the application might jeopardize the national interest, the Commissioner shall thereupon maintain the application in a sealed condition and notify the applicant thereof. The owner of an application which has been placed under a secrecy order shall have a right to appeal from the order to the Secretary of Commerce under rules prescribed by him.

An invention shall not be ordered kept secret and the grant of a patent withheld for a period of more than one year. The Commissioner shall renew the order at the end thereof, or at the end of any renewal period, for additional periods of one year upon notification by the head of the department or the chief officer of the agency who caused the order to be issued that an affirmative determination has been made that the national interest continues so to require. An order in effect, or issued, during a time when the United States is at war, shall remain in effect for the duration of hostilities and one year following cessation of hostilities. An order in effect, or issued, during a national emergency declared by the President shall remain in effect for the duration of the national emergency and six months thereafter. The Commissioner may rescind any order upon notification by the heads of the departments and the chief officers of the agencies who caused the order to be issued that the publication or disclosure of the invention is no longer deemed detrimental to the national security.

S182. Abandonment of invention for unauthorized disclosure
The invention disclosed in an application for patent subject to an order made pursuant to

section 181 of this title may be held abandoned upon its being established by the Commissioner that in violation of said order the invention has been published or disclosed or that an application for a patent therefor has been filed in a foreign country by the inventor, his successors, assigns, or legal representatives, or anyone in privity with him or them, without the consent of the Commissioner. The abandonment shall be held to have occurred as of the time of violation. The consent of the Commissioner shall not be given without the concurrence of the heads of the departments and the chief officers of the agencies who caused the order to be issued. A holding of abandonment shall constitute forfeiture by the applicant, his successors, assigns, or legal representatives, or anyone in privity with him or them, of all claims against the United States based upon such invention.

S183. Right to compensation

An applicant, his successors, assigns, or legal representatives, whose patent is withheld as herein provided, shall have the right, beginning at the date the applicant is notified that, except for such order, his application is otherwise in condition for allowance, or February 1, 1952, whichever is later, and ending six years after a patent is issued thereon, to apply to the head of any department or agency who caused the order to issued for compensation for the damage caused by the order of secrecy and/or for the use of the invention by the Government, resulting from his disclosure. The right to compensation for use shall begin on the date of the first use of the invention by the Government. The head of the department or agency is authorized, upon the presentation of a claim, to enter into an agreement with the applicant, his successors, assigns, or legal representatives, in full settlement for the damage and/or use. This settlement agreement shall be conclusive for all purposes notwithstanding any other provision of law to the contrary. If full settlement of the claim cannot be effected, the head of the department or agency may award and pay to such applicant, his successors, assigns, or legal representatives, a sum not exceeding 75 per centum of the sum which the head of the department or agency considers just compensation for the damage and/or use. A claimant may bring suit against the United States in the United States Claims Court or in the District Court of the United States for the district in which such claimant is a resident for an amount which when added to the award shall constitute just compensation for the damage and/or use of the invention by the Government. The owner of any patent issued upon an application that was subject to a secrecy order issued pursuant to section 181 of this title, who did not apply for compensation as above provided, shall have the right, after the date of issuance of such patent, to bring suit in the United States Claims Court for just compensation for the damage caused by reason of the order of secrecy and/or use by the Government of the invention resulting from his disclosure. The right to compensation for use shall begin on the date of the first use of the invention by the Government. In a suit under the provisions of this section the United States may avail itself of all defenses it may plead in and action under section 1498 of title 28. This section shall not confer a right of action on anyone or his successors, assigns, or legal representatives who, while in the full-time employment or service of the United States, discovered, invented, or developed the invention on which the claim is based.

S184. Filing of application in foreign country

Except when authorized by a license obtained from the Commissioner a person shall not file or cause or authorize to be filed in any foreign country prior to six months after filing in the United States an application for patent or for the registration of a utility model, industrial design, or model in respect of an invention made in this country. A license shall not be granted with respect to an invention subject to an order issued by the Commissioner pursuant to section 181 of this title without the concurrence of the head of the departments and the chief officers of the agencies who caused the order to be issued. The license may be granted retroactively where an application has been inadvertently filed abroad and the application does not disclose an invention within the scope of section 181 of this title.

The term "application" when used in this chapter includes applications and any modifications, amendments, or supplements thereto, or divisions thereof.

S185. Patent barred for filing without license
Notwithstanding any other provisions of law any person, and his successors, assigns, or legal representatives, shall not receive a United States patent for an invention if that person, or his successors, assigns, or legal representatives shall, without procuring the license prescribed in section 184 of this title, have made, or consented to or assisted another's making, application in a foreign country for a patent or for the registration of a utility model, industrial design, or model in respect of the invention. A United States patent issued to such person, his successors, assigns, or legal representatives shall be invalid.

S186. Penalty
Whoever, during the period or periods of time an invention has been ordered to be kept secret and the grant of a patent thereon withheld pursuant to section 181 of this title, shall, with knowledge of such order and without due authorization, willfully publish or disclose or authorize or cause to be published or disclosed the invention, or material information with respect thereto, or whoever, in violation of the provisions of section 184 of this title, shall file or cause or authorize to be filed in any foreign country an application for patent or for the registration of a utility model, industrial design, or model in respect of any invention made in the United States, shall, upon conviction, be fined not more than $10,000 or imprisoned for not more than two years, or both.

S187. Nonapplicability to certain persons
The prohibitions and penalties of this chapter shall not apply to any officer or agent of the United States acting within the scope of his authority, nor to any person acting upon his written instructions or permission.

S188. Rules and regulations, delegation of power
The Atomic Energy Commission, the Secretary of a defense department, the chief officer of any other department or agency of the Government designated by the President as a defense agency of the United States, and the Secretary of Commerce, may separately issue rules and regulations to enable the respective department or agency to carry out the provisions of this chapter, and may delegate any power conferred by this chapter.

Ownership and Assignment

S261. Ownership; assignment
Subject to the provisions of this title, patents shall have the attributes of personal property.

Applications for patent, patents, or any interest therein, shall be assignable in law by an instrument in writing. The applicant, patentee, or his assigns or legal representatives may in like manner grant and convey an exclusive right under his application for patent, or patents, to the whole or any specified part of the United States.

A certificate of acknowledgment under the hand and official seal of a person authorized to administer oaths within the United States, or, in a foreign country, of a diplomatic or consular officer of the United States or an officer authorized to administer oaths whose authority is proved by a certificate of a diplomatic or consular officer of the United States, or apostille of an official designated by a foreign country which, by treaty or convention, accords like effect to apostilles of designated officials in the United States, shall be prima facie evidence of the execution of an assignment, grant or conveyance of a patent or application for patent.

An assignment, grant or conveyance shall be void as against any subsequent purchaser or mortgagee for a valuable consideration, without notice, unless it is recorded in the Patent and Trademark Office within three months from its date or prior to the date of such subsequent purchase or mortgage.

S262. Joint owners
In the absence of any agreement to the contrary, each of the joint owners of a patent may make, use or sell the patented invention without the consent of and without accounting to the other owners.

Infringement of Patents

S271. Infringement of patent
(a) Except as otherwise provided in this title, whoever without authority makes, uses or sells any patented invention, within the United States during the term of the patent therefor, infringes the patent.

(b) Whoever actively induces infringement of a patent shall be liable as an infringer.

(c) Whoever sells a component of a patented machine, manufacture, combination or composition, or a material or apparatus for use in practicing a patented process, constituting a material part of the invention, knowing the same to be especially made or especially adapted for use in an infringement of such patent, and not a staple article or commodity of commerce suitable for substantial noninfringing use, shall be liable as a contributory infringer.

(d) No patent owner otherwise entitled to relief for infringement or contributory infringement of a patent shall be denied relief or deemed guilty of misuse or illegal extension of the patent right by reason of his having done one or more of the following: (1) derived revenue from acts which if performed by another without his consent would constitute contributory infringement of the patent; (2) licensed or authorized another to perform acts which if performed without his consent would constitute contributory infringement of the patent; (3) sought to enforce his patent rights against infringement or contributory infringement.

Remedies for Infringement of Patent, and Other Actions

S281. Remedy for infringement of patent
A patentee shall have remedy by civil action for infringement of his patent.

S282. Presumption of validity; defenses
A patent shall be presumed valid. Each claim of a patent (whether in independent or dependent form) shall be presumed valid independently of the validity of other claims; dependent claims shall be presumed valid even though dependent upon an invalid claim. The burden of establishing invalidity of a patent or any claim thereof shall rest on the party asserting it.

The following shall be defense in any action involving the validity or infringement of a patent and shall be pleaded:

(1) Noninfringement, absence of liability for infringement, or unenforceability,

(2) Invalidity of the patent or any claim in suit on any ground specified in part II of this title (i.e. 35 U.S.C. SS102 and 103) as a condition for patentability,

(3) Invalidity of the patent or any claim in suit for failure to comply with any requirements of sections 112 or 251 of this title.

(4) Any other fact or act made a defense by this title.

In actions involving the validity or infringement of a patent the party asserting invalidity or noninfringement shall give notice in the pleadings or otherwise in writing to the adverse party at least thirty days before the trial, or the country, number, date, and name of the patentee of any patent, the title, date, and page numbers of any publication to be relied upon as anticipation of the patent in suit, or except in actions in the United States Court of Claims, as showing the state of the art, and the name and address of any person who may be relied upon as the prior inventor or as having prior knowledge of or as having previously used or offered for sale the invention of the patent in suit. In the absence of such notice proof of the said matters may not be made at the trial except on such terms as the court requires.

S283. Injunction
The several courts having jurisdiction of cases under this title may grant injunctions in accordance with the principles of equity to prevent the violation of any right secured by patent, on such terms as the court deems reasonable.

S284. Damages
Upon finding for the claimant the court shall award the claimant damages adequate to compensate for the infringement but in no event less than a reasonable royalty for the use made of the invention by the infringer, together with interests and costs as fixed by the court.

When the damages are not found by a jury, the court shall assess them. In either event the court may increase the damages up to three times the amount found or assessed.

The court may receive expert testimony as an aid to the determination of damages or of what royalty would be reasonable under the circumstances.

S285. Attorney fees
The court in exceptional cases may award reasonable attorney fees to the prevailing party.

S286. Time limitation on damages
Except as otherwise provided by law, no recovery shall be had for any infringement committed more than six years prior to the filing of the complaint or counterclaim for infringement in the action.

In the case of claims against the United States Government for use of a patented invention, the period before bringing suit, up to six years, between the date of receipt of a written claim for compensation by the department or agency of the Government having authority to settle such claim, and the date of mailing by the Government of a notice to the claimant that his claim has been denied shall not be counted as part of the period referred to in the preceding paragraph.

S287. <u>Limitation on damages; marking and notice</u>
Patentees, and persons making or selling any patented article for or under them, may give notice to the public that the same is patented, either by fixing thereon the word "patent" or the abbreviation "pat.", together with the number of the patent, or when, from the character of the article, this can not be done, by fixing to it, or to the package wherein one or more of them is contained, a label containing a like notice. In the event of failure so to mark, no damages shall be recovered by the patentee in any action for infringement, except on proof that the infringer was notified of the infringement and continued to infringe thereafter, in which event damages may be recovered only for infringement occurring after such notice. Filing of an action for infringement shall constitute such notice.

S288. <u>Action for infringement of a patent containing an invalid claim</u>
Whenever, without deceptive intention, a claim of a patent is invalid, an action may be maintained for the infringement of a claim of the patent which may be valid. The patentee shall recover no costs unless a disclaimer of the invalid claim has been entered at the Patent and Trademark Office before the commencement of the suit.

S292. <u>False marking</u>
(a) Whoever, without the consent of the patentee, marks upon, or affixes to, or uses in advertising in connection with anything made, used, or sold by him, the name or any imitation of the name of the patentee, the patent number, or the words "patent," "patentee," or the like, with the intent of counterfeiting or imitating the mark of the patentee, or of deceiving the public and inducing them to believe that the thing was made or sold by or with the consent of the patentee; or

Whoever marks upon, or affixes to, or uses in advertising in connection with any unpatented article, the word "patent" or any word or number importing the same is patented, for the purpose of deceiving the public; or

Whoever marks upon, or affixes to, or uses in advertising in connection with any article, the words "patent applied for," "patent pending," or any word importing that an application for patent has been made, when no application for patent has been made, or if made, is not pending, for the purpose of deceiving the public --

Shall be fined not more than $500 for every such offense. (b) Any person may sue for the penalty, in which event one-half shall go to the person suing and the other to the use of the United States.

S294. <u>Voluntary arbitration</u>
(a) A contract involving a patent or any right under a patent may contain a provision requiring arbitration of any dispute relating to patent validity or infringement arising under the contract. In the absence of such a provision, the parties to an existing patent validity or infringement dispute may agree in writing to settle such dispute by arbitration. Any such provision or agreement shall be valid, irrevocable, and enforceable, except for any grounds that exist at law or in equity for revocation of a contract.

(b) Arbitration of such disputes, awards by arbitrators and confirmation of awards shall be governed by title 9, United States Code, to the extent such title is not inconsistent with this section. In any such arbitration proceeding, the defenses provided for under section 282 of this title shall be considered by the arbitrator if raised by any party to the proceeding.

(c) An award by an arbitrator shall be final and binding between the parties to the arbitration but shall have no force or effect on any other person. The parties to an arbitration may agree that in the event of a patent which is the subject matter of an award is subsequently determined to be invalid or unenforceable in a judgment rendered by a court to competent jurisdiction from which no appeal can or has been taken, such award may be modified by any court of competent jurisdiction upon application by any party to the arbitration. Any such modification shall govern the rights and obligations between such parties from the date of such modification.

(d) When an award is made by an arbitrator, the patentee, his assignee or licensee shall give notice thereof in writing to the Commissioner. There shall be a separate notice prepared for each patent involved in such proceeding. Such notice shall set forth the names and addresses of the parties, the name of the inventor, and the name of the patent owner, shall designate the number of the patent, and shall contain a copy of the award. If an award is modified by a court, the party requesting such modification shall give notice of such modification to the Commissioner. The Commissioner shall, upon receipt of either notice, enter the same in the record of the prosecution of such patent. If the required notice is not filed with the Commissioner, any party to the proceeding may provide such notice to the Commissioner.

(e) The award shall be unenforceable until the notice required by subsection (d) is received by the Commissioner.

APPENDIX E: HOW TO ORDER ADDITIONAL COPIES OF THIS BOOK

To order additional copies of *Protecting Engineering Ideas & Inventions* write:

Penn Institute, Inc.
P.O. Box 41016
Cleveland, Ohio 44141

Or call toll-free 1-800-426-7495. (In the Cleveland area please call 237-2345.) Or Telex 276809.

Protecting Engineering Ideas & Inventions is available in paper back (ISBN 0-944606-01-6) and hard back (ISBN 0-944606-02-4). We accept checks, MasterCard, Visa, American Express, or purchase orders. Quantity discounts are available.

OCT 24 1993